Material and Energy Engineering for Sustainable Advancement (MEESA 2021)

AA007673

IOP Conference Series: Materials Science and Engineering
Volume 1217

Kota Kinabalu, Malaysia
29 - 30 September 2021

Editors:

Wan Khairul Muzammil bin Abd Rahim
Nancy Julius Siambun
Melvin Gan Jet Hong

ISBN: 978-1-7138-4676-5
ISSN: 1757-8981

Printed from e-media with permission by:

Curran Associates, Inc.
57 Morehouse Lane
Red Hook, NY 12571

proceedings.com

Some format issues inherent in the e-media version may also appear in this print version.

This work is licensed under a Creative Commons Attribution 3.0 International Licence.
Licence details: http://creativecommons.org/licenses/by/3.0/.

No changes have been made to the content of these proceedings. There may be changes to pagination and minor adjustments for aesthetics.

Printed with permission by Curran Associates, Inc. (2022)

For permission requests, please contact the Institute of Physics at the address below.

Institute of Physics
Dirac House, Temple Back
Bristol BS1 6BE UK

Phone: 44 1 17 929 7481
Fax: 44 1 17 920 0979

techtracking@iop.org

Additional copies of this publication are available from:

Curran Associates, Inc.
57 Morehouse Lane
Red Hook, NY 12571 USA
Phone: 845-758-0400
Fax: 845-758-2633
Email: curran@proceedings.com
Web: www.proceedings.com

TABLE OF CONTENTS

IOP Conf. Series: Materials Science and Engineering 1217 (2022) 011001 doi:10.1088/1757-899X/1217/1/011001

MATERIAL AND ENERGY ENGINEERING FOR SUSTAINABLE ADVANCEMENT (MEESA 2021) CONFERENCE

Material and Energy Engineering for Sustainable Advancement

29th – 30th SEPTEMBER 2021

ORGANIZER

MECHANICAL ENGINEERING
FACULTY OF ENGINEERING
UNIVERSITI MALAYSIA SABAH

Content from this work may be used under the terms of the Creative Commons Attribution 3.0 licence. Any further distribution of this work must maintain attribution to the author(s) and the title of the work, journal citation and DOI.
Published under licence by IOP Publishing Ltd

IOP Conf. Series: Materials Science and Engineering 1217 (2022) 011001 doi:10.1088/1757-899X/1217/1/011001

Preface

The Material & Energy Engineering for Sustainable Advancement 2021 (MEESA 2021) conference was organized by the Mechanical Engineering Programme, Faculty of Engineering, Universiti Malaysia Sabah. The conference was held virtually on the 29th and 30th of September 2021.

The 1st MEESA conference was organized to enculturate research and collaborative involvement between researchers in the engineering materials and energy field. Researchers and postgraduate students from multidisciplinary research backgrounds gathered to share new findings and ultimately building a better network for long term collaboration opportunities. The silver lining in organizing this online conference is the participation of international researchers, offering a more diverse network to be built between participants. Twenty-one papers have been presented in MEESA 2021, of which a total of 17 papers was selected for this publication. Four keynote speakers, who are experts in their respective fields were invited.

The keynote speakers were Dr. Habibah binti Ghazali (Universiti Teknologi Malaysia), Dr. Ong Hwai Chyuan (University of Technology Sydney), Ir. Dr. Mohd Azlan bin Ismail (Universiti Malaysia Sabah) and Ir. Ts. Dr. Melvin Gan Jet Hong (Universiti Malaysia Sabah). Although this year's conference was held virtually, we strongly believe that this conference does not reduce the purpose or benefit to the scientific world,in particular, and the general public to share current issues and the latest findings in the field of materials and energy for our environment. All manuscripts published in this proceeding have been vetted through a rigorous review to meet the requirement of high-quality papers.

Lastly, we would like to thank the MEESA 2021 Advisor: Dean of Faculty of Engineering, Assoc. Prof. Ts. Dr. Ismail bin Saad, session chairpersons, and members of the Scientific and Technical Committees for their support and undivided commitment to make the conference possible. The committee would also wish to acknowledge all the keynote speakers, reviewers and all participants for their time and effort in making this conference a success.

Warmest Regards,

Organizing Committee of the 1st MEESA 2021

IOP Conf. Series: Materials Science and Engineering 1217 (2022) 011001 doi:10.1088/1757-899X/1217/1/011001

ORGANIZING COMMITTE

ADVISOR
Assoc. Prof. Ts Dr. Ismail Bin Saad

CHAIRMAN
Ir. Dr. Chua Bih Lii

DEPUTY CHAIRMAN
Assoc. Prof. Ts. Dr. Nancy Julius Siambun

SECRETARIAT
Dr. Nazrein Adrian bin Amaludin
Ir. Ts. Dr. Melvin Gan Jet Hong

TREASURER
Mdm. Fadzlita Mohd Tamiri

EDITORS (IOP PROCEEDING)
Dr. Wan Khairul Muzammil bin Abd Rahim (Chair Editor)
Assoc. Prof. Ts. Dr. Nancy Julius Siambun
Ir. Ts. Dr. Melvin Gan Jet Hong

SCIENTIFIC COMMITTEE
Prof. Dr. Willey Liew Yun Hsein
Dr. Wan Khairul Muzammil bin Abd Rahim
Ir. Dr. Mohd Azlan Ismail
Ts. Dr. Farm Yan Yan
Dr. Choong Wai Heng

TECHNICAL COMMITTEE
Dr. Mohd Kamel Wan Ibrahim
Dr. Mohd Suffian Misaran
Dr. Noor Ajian Mohd Lair
Dr. Mohd Kamal Mohd Shah
Mr. Abdullah Mohd Tahir
Ir. Ts. Mohd Adzrie Radzali
Ir. Ts. Mohd Amran bin Hj. Madlan

IOP Conf. Series: Materials Science and Engineering 1217 (2022) 011002 doi:10.1088/1757-899X/1217/1/011002

Peer review declaration

All papers published in this volume of *IOP Conference Series: Materials Science and Engineering* have been peer reviewed through processes administered by the Editors. Reviews were conducted by expert referees to the professional and scientific standards expected of a proceedings journal published by IOP Publishing.

- **Type of peer review: Single-blind**
  The single-blind peer-review method was used for the peer-review process.

- **Conference submission management system: The papers are emailed to the Secretariat and managed internally.**

- **Number of submissions received: 21**

- **Number of submissions sent for review: 21**

- **Number of submissions accepted: 17**

- **Acceptance Rate (Number of Submissions Accepted / Number of Submissions Received X 100): 81%**

- **Average number of reviews per paper: 2**

- **Total number of reviewers involved: 26**

- **Any additional info on review process:**

  Secretariat stage

  After receiving a new submission, the Secretariat conducts initial pre-refereeing checks to ensure the article is legible, complete, in the style of a scientific paper, written in clear English, and within the scope of the conference. Articles that do not possess these criteria would be rejected at this stage.

  Scientific Committee stage

  Manuscripts that the Secretariat has accepted would then be given to the Scientific Committee for similarity checking using Turnitin. Articles with a higher than 30% similarity will be rejected, whereas those with similarities below than 30% will be accepted and processed to begin a formal peer review.

  Research papers submitted for the conference are sent to two independent referees who are asked to report on the following:

  Technical Criteria
  - Scientific merit: notably scientific rigour, accuracy, and correctness.

Content from this work may be used under the terms of the Creative Commons Attribution 3.0 licence. Any further distribution of this work must maintain attribution to the author(s) and the title of the work, journal citation and DOI.
Published under licence by IOP Publishing Ltd

- Does the manuscript demonstrate an adequate understanding of the relevant literature in the field?
- Does the manuscript demonstrate sufficient discussion of the context of the work and suitable referencing?

Quality Criteria

- Is the work relevant and novel?
- Does the problem considered have a sound motivation?
- Is the content of the work of sufficient scientific interest to justify its length?

Presentation Criteria

- Is the title adequate and appropriate for the content of the article?
- Does the abstract contain complete and essential information about the article?
- Are the diagrams, figures, tables, and captions essential and clear?
- Does the paper contain a carefully written and clear conclusion?

Referees are chosen from the conference organiser's reviewer database. The Scientific Committee would find the best combination of scientific expertise and referee experience for each paper submitted to the conference.

The authors are welcome to revise their papers if any referees recommend a revision. Once the correction has been made, the Scientific Committee reviews the revised paper and the referee's reports. Then the Scientific Committee distributes the revised paper back to the referees for final confirmation.

Adjudicator

If the referees are not in agreement, the paper and the reports are sent to an independent adjudicator (in this case, a member of the Scientific Committee) who will be tasked to review the article and read the referee's reports and adjudicate between them. Then, a formal decision will be made based on the adjudicator's recommendation, and all referees will be notified.

- **Contact person for queries:**
  1. Dr Wan Khairul Muzammil bin Abdul Rahim (khairulm@ums.edu.my)
  2. Dr Nazrein Adrian Amaludin (nazrein.amaludin@ums.edu.my)

Desalination of seawater using carbon-coated solar absorber in solar still

W M J Karen[1,*], C Y Wong[1], G J H Melvin[2,*]

1 Faculty of Engineering, Universiti Malaysia Sabah, Jalan UMS, 88400 Kota Kinabalu, Sabah, Malaysia.
2 Centre of Research in Energy and Advanced Materials, Faculty of Engineering, Universiti Malaysia Sabah, Jalan UMS, 88400 Kota Kinabalu, Sabah, Malaysia.

*Corresponding authors: dk1911006t@student.ums.edu.my, melvin.gan@ums.edu.my

Abstract. Desalination is a process of eliminating salt and other minerals from seawater which turns it to a safe usable water. This study showed the generation of clean water from seawater using carbon-coated solar absorber in two different solar still body colours which are transparent and black solar still, under direct solar exposure at approximately 1.2 kW/m^2. The efficiency of carbon-coated solar absorber in the transparent and black solar still was computed, while the clean water collected was examined for its pH and salinity. Carbon-coated solar absorber in black solar still exhibit the highest efficiency at around 35.71%, where the pH and salinity of the collected clean water was substantially reduced to usable water at 6.55 and 62 ppm, respectively.

1. Introduction

According to the World Health Organization in 2019 [1], about half of the population in the world will live in water deficiency regions by the year of 2025. One of the solutions in tackling the problem is by generating clean water from seawater. Low-cost and ecofriendly solar desalination utilizes solar energy to transform seawater into clean water in a solar still [2]. Conventional solar desalination generated low yield of clean water as high energy was required to heat up the water body. Optimization on the solar desalination was studied by, (i) focusing on minimizing the heat loss with heat localization on the photothermal conversion material surface, and (ii) specific water pathways from the water body to the evaporation surface [3, 4].

Carbon-based materials are an excellent photothermal conversion material for solar desalination as they have a broad light absorption capability, great stability and thermal conductivity [5–7]. Furthermore, low-cost carbon materials can be produced from agricultural waste such as rice husk. Rice husk is a common waste from the rice industry, and it is usually burned or thrown away. However, it can be converted to valuable carbon material by heat treatment, where the high carbon content of the carbonized rice husk could be utilized for various applications [8, 9]. Porous structure and black colour of the carbonized rice husk is highly advantageous since the attributes can channel water and absorb solar radiation to generate heat [2, 10]. Fang and associates [10, 11] worked on the development of carbonized rice husk-based foam for solar vapor generation. They found that the carbonized rice husk-based solar absorber possesses excellent solar adsorption ability and hydrophilicity due to its porous structures which leads to good conversion efficiency around 71% at 1 sun intensity in deionized water. Further improvement on the carbonized rice husk-based foam was done by addition of plasmonic material,

Content from this work may be used under the terms of the Creative Commons Attribution 3.0 licence. Any further distribution of this work must maintain attribution to the author(s) and the title of the work, journal citation and DOI.
Published under licence by IOP Publishing Ltd

which was gold nanoparticles. The enhanced carbonized rice husk foam with gold nanoparticles displayed an increase in conversion efficiency at about 84% in deionized water, whereas slightly lower efficiency was observed (~75%) in 15 wt % of NaCl solution, where both setups were tested at 1 sun intensity.

Development on the water pathways from the water body to the evaporation surface showed a favourable outcome. Application of air gap and capillary between the water body and solar absorber reduced the heat loss via conduction to the water body [12, 13]. Furthermore, the capillary constantly provides water to the evaporation surface as a hydrophilic material was usually chosen such as cellulose sponge [14], cotton towel [15], and paper [16].

The enhancement on the clean water generation in a solar still was studied by implementing carbon based solar absorber to localized heat directly onto the solar absorber surface rather than heating up the seawater body. There were limited studies which utilized carbonized rice husk as the photothermal conversion material for clean water generation from seawater. Two different solar still body colours, transparent and black, were fabricated and clean water was generated from seawater by solar driven desalination. Waste rice husk was used in this study as the source of the carbon material, and melamine sponge was utilized to channel the seawater as it possesses excellent capillarity. The efficiency of the solar absorber along with the salinity and pH values was observed.

2. Experimental

Fabrication of solar stills and the experimental procedure was done based on reference [17] with some modification. Solar stills were constructed using 3 mm thick non-reflective transparent and black acrylic plastic, respectively. The dimension of the solar stills was about 150 x 200 x 180 mm (width x length x height), with transparent top cover at 16° inclination for both transparent and black solar stills, which the solar radiation entered through the top cover. A 13 mm thick polystyrene foam was used as the thermal insulator, while cotton cloth was utilized as the capillary to transport seawater. Figure 1 displays the schematic diagram of clean water production from seawater using carbon-coated solar absorber in the solar still.

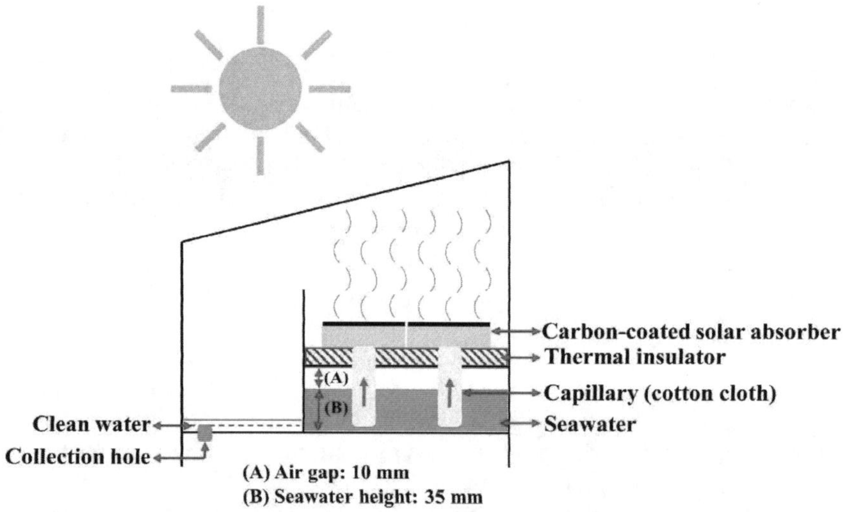

Figure 1. Schematic diagram of clean water production from seawater using carbon-coated solar absorber in the solar still.

Rice husk (Fatt Agriculture Sdn. Bhd., Keningau) was carbonized using a furnace in an argon gas environment for two hours at 500°C, naturally cooled down and was crushed to powder form. Carbonized rice husk (CRH), polyvinyl alcohol (PVA, Aldrich Chemistry) and ethanol (EtOH, Fisher

Scientific (M) Sdn. Bhd.) were mixed in a weight ratio of 0.001:1:2 (PVA:CRH:EtOH) and sonicated to achieve a homogenous mixture. The homogenous mixture was coated (1 mm coating thickness) at the top surface of the melamine sponge with dimension of 120 x 100 x 25 mm (height x width x thickness).

Table 1 shows the experimental setup of the transparent and black solar still, with bulk seawater and carbon-coated solar absorber. The weight of seawater used in all setups were the same, where about 500 g of seawater was filled in the solar still basin with the height of approximately 35 mm, and 10 mm of air gap between the thermal insulator and seawater body. The solar stills were left under the sun for approximately 4 hours, and the solar intensity (Solar Power Meter SM206-SOLAR) including the top surface temperature (Infrared Digital Thermometer GM320) of bulk seawater and carbon-coated solar absorber were taken at 15 minutes intervals. Clean water was collected at the end of the experiment and its mass was recorded (Digital Scale I-2000, 500 ± 0.01g).

Table 1. Experimental setup of the transparent and black solar still.

Experiment	Solar still body colour	Setup	Solar absorber dimension (mm) Height x width x thickness
TS	Transparent	Bulk seawater	-
BS	Black	Bulk seawater	-
TC	Transparent	Carbon-coated solar absorber	120 x 100 x 25
BC	Black	Carbon-coated solar absorber	120 x 100 x 25

Efficiency, η (%), of the bulk seawater and carbon-coated solar absorber was calculated using the equation below.

$$\eta = \frac{\dot{m}\Delta H_{LV}}{I} \quad (1)$$

Where $\dot{m}$ is the water mass flux under solar radiation (kg/m^2 h), ΔH_{LV} is the enthalpy change due to water evaporation (kJ/kg), and I is the solar intensity (kW/m^2).

Salinity and pH values of the seawater (initial) and collected clean water (final) were taken to observe the quality of the clean water, where it was measured by a water quality meter (Water Quality Tester EZ-9901).

3. Results and discussion

The transparent and black solar stills setup during the direct exposure to solar radiation at the beginning, and end of the experiment, as well as the thermal images of the top solar stills during the experiment is shown in figure 2. After about 4 hours, water droplets were observed at the top cover of the solar stills as displayed in figure 2 (b), since the water vapor condensed when it encountered the cold top cover. The mass of the clean water generated was about 8.22 g, 13.97 g, 28.55 g, and 32.66 g for TS, BS, TC and BC respectively. Greater yield of clean water was obtained for TC and BC as the evaporation process at the top surface of the carbon-coated solar absorber required less energy and time, compared to the conventional evaporation process for TS and BS. Thermal images captured from the top of the solar stills in figure 2 (c) shows that the black solar still, BS and BC, efficiently absorb and trap heat, compared to the transparent solar still, TS and TC. Black colour is known to effectively absorb light rather than reflect it, thus the black solar still converts more heat.

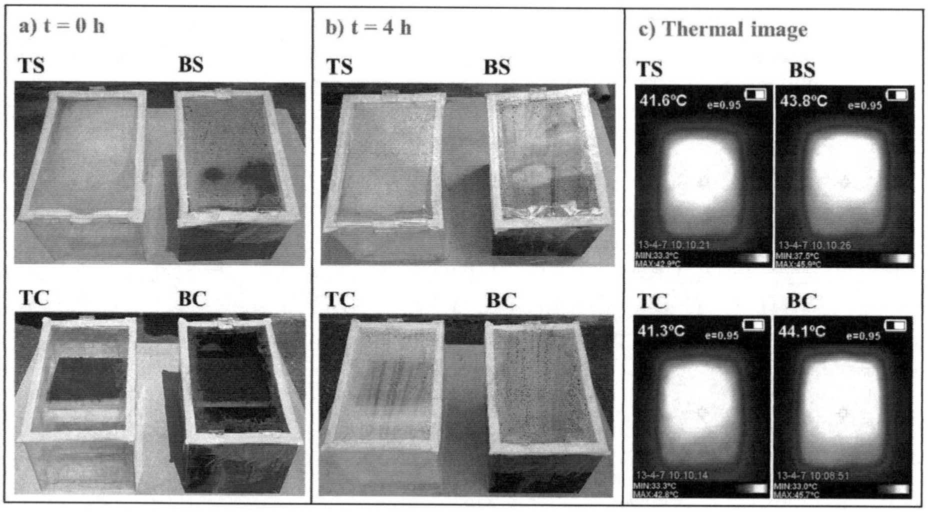

Figure 2. The setup at (a) the beginning (t = 0 h), (b) end of the experiment (t = 4 h), and (c) thermal images of the top solar stills.

Figure 3 indicated the top surface temperature of bulk seawater and solar absorber along with the solar intensity throughout the experiment, and the efficiency of TS, BS, TC and BC. Based on figure 3 (a), BC (30.9°C ~ 53.7°C) displayed the highest top surface temperature, while TC (30.7°C ~ 48.8°C) and BS (30.9°C ~ 48.6°C) showed almost similar trend, and TS (30.9°C ~ 41.3°C) exhibited low range of top surface temperature. The generated heat was concentrated at the top surface of the carbon-coated solar absorber, therefore higher top surface temperature was observed for TC and BC. Without the aid of carbon-coated solar absorber in TS and BS, the heat generated in the solar still was absorbed by the bulk seawater. Variation in the solar radiation intensity at approximately 1.02 ~ 1.39 kW/m^2 because the experiment was done outdoors. The average solar intensity was observed at about 1.2kW/m^2. Efficiency of TS, BS, TC and BC (figure 3 (b)) was calculated at 3.70%, 8.28%, 30.76%, and 35.71%, respectively.

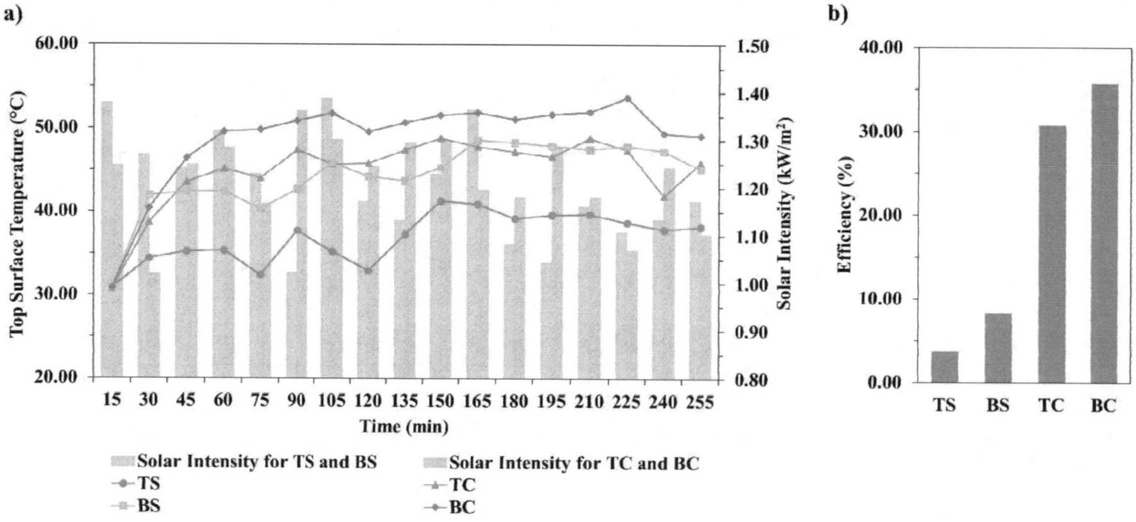

Figure 3. The (a) top surface temperature and solar intensity during the experiment, and (b) efficiency of TS, BS, TC, and BC.

Porous structure and black colour of the carbonized rice husk effectively absorbs the solar radiation and generates heat, hence TC and BC showed high top surface temperature in contrast to TS. However, the black colour body of the solar still contributes to the elevated top surface temperature for BS. Higher efficiency of TC and BC was due to the superb capillarity of the porous carbonized rice husk and melamine sponge, where it continuously contained and supplied seawater for the evaporation process without heat loss to the seawater body [2, 10, 18]. Moreover, the restricted water pathway using the cotton cloth as the capillary further aids in minimizing the heat loss. The air gap between the seawater body and solar absorber creates a border and separates the two components while seawater is continuously supplied to the top surface via the capillary [12–16]. Implementation of carbon-coated solar absorber and restricted water pathway for BC, in addition to the black colour body of the solar still consequently able to maintain the heat inside the solar still and generated a great yield of clean water.

Salinity and pH values of the initial seawater and collected clean water was presented in table 2. The collected clean water salinity was recorded at 224 ppm, 216 ppm, 58 ppm and 62 ppm for TS, BS, TC and BC respectively, where a reduction of about 98.7 ~ 99.7% was observed from the initial seawater salinity at around 17560 ppm. The presence of salt (salinity) in the collected clean water was because of the evaporation process in the closed solar still. As the closed solar still was concentrated with the water vapor, the molecules regularly bumped onto the top surface of the bulk seawater and carbon-coated solar absorber, hence giving weak kinetic energy for the salt molecules to escape [17]. Besides, porous structure of melamine sponge and carbonized rice husk helps in filtering the salt molecules, therefore lower salinity level in TC and BC [18]. Whereas the pH value of the clean water was roughly similar which was between 6.55 and 6.67. Based on the guidelines by World Health Organization [19,20], the salinity level of drinkable water is 500 ppm, while the pH is between 6.5 and 8.5. Thus, the salinity and pH values of the clean water falls in the drinkable water level.

Table 2. Salinity and pH value of seawater (initial) and collected clean water (final).

Experiment	Salinity (ppm)		pH	
	Initial	Final	Initial	Final
TS	17560	224	7.84	6.65
BS	17560	216	7.84	6.67
TC	17560	58	7.84	6.54
BC	17560	62	7.84	6.55

4. Conclusion

Clean water was successfully generated from both transparent and black solar stills with bulk seawater and carbon-coated solar absorber, under direct solar radiation at about 1.2 kW/m^2 for around 4 hours. The black body colour of solar still proficiently absorbs and traps the heat inside the solar still. Application of carbon-coated solar absorber and restricted water pathway in the enhanced setup exhibited tremendous improvement in the generation and entrapment of heat, hence showed higher efficiency. Therefore, the black solar still with carbon-coated solar absorber displays the highest efficiency at 35.71%, with salinity and pH within the drinkable water level at 62 ppm and 6.55, respectively.

Acknowledgement
This work was supported by Universiti Malaysia Sabah (Grant number GUG0386-2/2019), Malaysia.

IOP Conf. Series: Materials Science and Engineering 1217 (2022) 012001 doi:10.1088/1757-899X/1217/1/012001

References

[1] World Health Organization. Drinking-water [Internet]. 2019 [updated 2019 June 14; cited 2021 August 20]. Available from: https://www.who.int/news-room/fact-sheets/detail/drinking-water

[2] Ding T, Zhou Y, Ong W L and Ho G W 2021 *Mater. Today* **42** 178–91

[3] Zhou Y, Ding T, Gao M, Chan K H, Cheng Y, He J and Ho G W 2020 *Nano Energy* **77** 105102

[4] Gao M, Peh C K, Zhu L, Yilmaz G and Ho G W 2020 *Adv. Energy Mater.* **10** 1–9

[5] Min X, Zhu B, Li B, Li J and Zhu J 2021 *Accounts Mater. Res.* **2** 198–209

[6] Li Z, Ma W, Xu J, Qian M, Bi H, Zhao W, Lv Z and Huang F 2020 *Carbon* **170** 256–63

[7] He W, Zhou L, Wang M, Cao Y, Chen X and Hou X 2021 *Sci. Bull.* **66** 1472–83

[8] Melvin G J H, Chai K F and Tamiri F M 2019 *IOP Conf. Ser. Mater. Sci. Eng.* **606** 012002

[9] Karen W M J, Ooi J X, Wang Z, Siambun N J, Liew W Y H, Muzammil W K, Sopian K and Melvin G J H 2021 *Int. J. Surf. Sci. Eng.* **15** 36–49

[10] Fang W, Zhao L, He X, Chen H, Li W, Zeng X, Chen X, Shen Y and Zhang W 2020 *Renew. Energy* **151** 1067–75

[11] Fang W, Chen H, He X, Li W, Zhang W, Shen Y, Chen X and Zhao L 2021 *J. Porous Mater.* https://doi.org/10.1007/s10934-021-01112-1

[12] Mu X, Gu Y, Wang P, Wei A, Tian Y, Zhou J, Chen Y, Zhang J, Sun Z, Liu J, Sun L, Tanemura S and Miao L 2021 *Sol. Energy Mater. Sol. Cells* **220** 110842

[13] Gao M, Zhu L, Peh C K and Ho G W 2019 *Energy Environ. Sci.* **12** 841–64

[14] Wu D, Du C and Huang C 2021 *Appl. Therm. Eng.* **195** 117238

[15] Yang Y, Sui Y, Cai Z and Xu B 2019 *Glob. Challenges* **3** 1900004

[16] Gong F (Frank), Li H, Wang W, Huang J, Xia D (David), Liao J, Wu M and Papavassiliou D V. 2019 *Nano Energy* **58** 322–30

[17] Karen W M J, Wong C Y, Wang Z, Liew W Y H, and Melvin G J H 2021 *Environ. Technol.* https://doi.org/10.1080/09593330.2021.1970820

[18] Liu P F, Miao L, Deng Z, Zhou J, Gu Y, Chen S, Cai H, Sun L and Tanemura S 2019 *Appl. Energy* **241** 652–9

[19] World Health Organization. pH in drinking-water, WHO Guidelines for Drinking-Water Quality [Internet]. 2007 [updated 2007; cited 2021 August 20]. Available from: https://www.who.int/water_sanitation_health/dwq/chemicals/ph_revised_2007_clean_version.pdf

[20] World Health Organization. Safe Drinking-water from Desalination [Internet]. 2011 [updated 2011; cited 2021 August 20]. Available from: http://apps.who.int/iris/bitstream/10665/70621/1/WHO_HSE_WSH_11.03_eng.pdf

IOP Conf. Series: Materials Science and Engineering 1217 (2022) 012002 doi:10.1088/1757-899X/1217/1/012002

Design of thermoformable three dimensional-printed PLA cast for fractured wrist

N P Sorimpuk*, W H Choong and B L Chua*

Faculty of Engineering, Universiti Malaysia Sabah, Jalan UMS, 88400 Kota Kinabalu, Sabah, Malaysia

*Corresponding author: neilson@polikk.edu.my; bihlii@ums.edu.my

Abstract. Patient specific plastic cast for broken limbs has been developed recently in pharmaceutical field through three-dimensional (3D) printing method. However, the production of a 3D printed cast through normal 3D printing method is time consuming compared to conventional plaster casting. In this study, a design of ventilated structured thermoformable 3D-printed polylactic acid (PLA) cast was produced as an alternative for the 3D printed cast production method. This design was initially printed in a flat shape and then transformed into a cast which can be fitted to the user's arm by using heat and external force. Finite Element Analysis (FEA) method was used to assess the mechanical properties of the proposed cast. In this analysis, the thermoformable design was exerted with a distributed force of 400 N, which is larger than the loading conditions encountered by human in their daily life. The mechanical properties of the thermoformable PLA cast such as local displacement under a specific load, maximum load, and stress were evaluated. Results were compared with the mechanical properties of Plaster of Paris cast. The results obtained from the FEA indicates that at the same layer thickness, the thermoformable 3D-printed PLA cast is stronger than the Plaster of Paris cast.

1. Background

A three-dimensional (3D) printed cast is a plastic shell protecting and immobilizing fractured or broken limb. Developed through the application of 3D scanner with advanced 3D modelling software and rapid prototyping machine, 3D printing enables patient specific cast and splint to be produced with high accuracy [1]. This product is now commonly developed to replace the current application of conventional Plaster of Paris (POP) cast. Several designs of plastic casts and splints have been developed through 3D printing method namely material extrusion (ME) process. Chen *et al.* concluded that 3D printed cast application has the advantage on the broken bone healing process based on their clinical study on a group of 10 patients [2].

The current techniques to produce a 3D printed arm cast is either through Fused Deposition Modelling (FDM) or Selective Laser Sintering (SLS) process, which are time and cost consuming. It is even much slower than traditional Plaster of Paris (POP) casting method [1]. The ventilated structure of the cast requires printing support. This leads to extra waste of material and construction time. The design flexibility of the cast will be limited as there needs to be hand/tool access to manually remove the support structure [3]. A simple 3D printed cast produced by Li *et al.* [4] consumed approximately ten hours of printing productions. Orthopaedic surgeons or medical practitioners involved with the 3D printed cast need to be exposed with knowledge in Computer Aided Design (CAD) and utilization of 3D Scanner in

Content from this work may be used under the terms of the Creative Commons Attribution 3.0 licence. Any further distribution of this work must maintain attribution to the author(s) and the title of the work, journal citation and DOI.
Published under licence by IOP Publishing Ltd

order to utilize the cast printing [4]. The cast produced by 3D scanning method is based on the imaging data of the limb during inflammatory phase. By the time the cast is produced, the swelling has already subsided and resulting in a cast that is not properly fit [5]. In FDM, casts are usually printed vertically and affected by anisotropic characteristics of 3D printing. Molten filament is deposited layer by layer in a crisscross manner and once it is extruded, the heat dissipates rapidly due to the colder ambient temperature. When the next layer is deposited, the first layer has cooled and therefore diffusion between the two layers is weakened. Therefore, the toughness and robustness of the printed structure is greatly affected [6].

Thermoforming is a general term that refers to the process of transforming a thermoplastic plastic sheet into a 3-dimensional shape by using heat [7]. In this study, thermoforming process is incorporated for a 3D printed cast to form the desirable 3D shape in order to shorten the cast production time. The cast was initially 3D printed as a flat object, heated at 60 ºC, and then transformed into a functional orthopaedic cast by wrapping it around the user's hand while it is still soft. The cast will harden itself as it cools off. Polylactic acid (PLA) was used to develop the flat 3D printed cast. The aim of this study is to investigate the mechanical properties of the proposed thermoformed 3D printed cast design by using finite element analysis method. The hypothesis is the thermoformed cast produced should be equal or stronger than the conventional plaster cast at the equivalent thickness.

2. Material and methods

2.1. Material

The material used in this study is polylactic acid (PLA) due to its due to its biodegradability and biocompatibility to human body [8]. The glass transition temperature of PLA with a molecular weight of 22,000 g mol^{-1} is 55 °C which means, it will starts softening at the temperature [9]. PLA can also be reshaped when heated several times without any change in their chemical or mechanical properties. Table 1 is the material's technical specification data provided by the manufacturer of the material.

Table 1. PLA Filament Technical Data Sheet by Ultimaker [10].

	Typical value	Test method
Brand/Manufacturer	Ultimaker	-
Diameter	2.85 ± 0.10 mm	-
Tensile modulus	2,346.5 MPa	ISO 527 (1 mm/min)
Tensile stress at yield	49.5 MPa	ISO 527 (50 mm/min)
Tensile stress at break	45.6 MPa	ISO 527 (50 mm/min)
Elongation at yield	3.3%	ISO 527 (50 mm/min)
Elongation at break	5.2%	ISO 527 (50 mm/min)
Flexural strength	103 MPa	ISO 178
Flexural modulus	3,150 MPa	ISO 178

2.2. Design

The orthopaedic casting concept for this design is unique; it was initially designed and fabricated as a flat shaped plastic sheet, heated until soften and then wrapped around the user's hand. The cast will then solidify into a rigid structure as it cools down. The cast design was developed by considering four major characteristics; practical design, opening holes for ventilation, the cast thermoformability, and the constraint mechanism. CATIA V5R20 (Dassault System, France) software was used as the main software for the design development. The cast design concept is illustrated in figure 1.

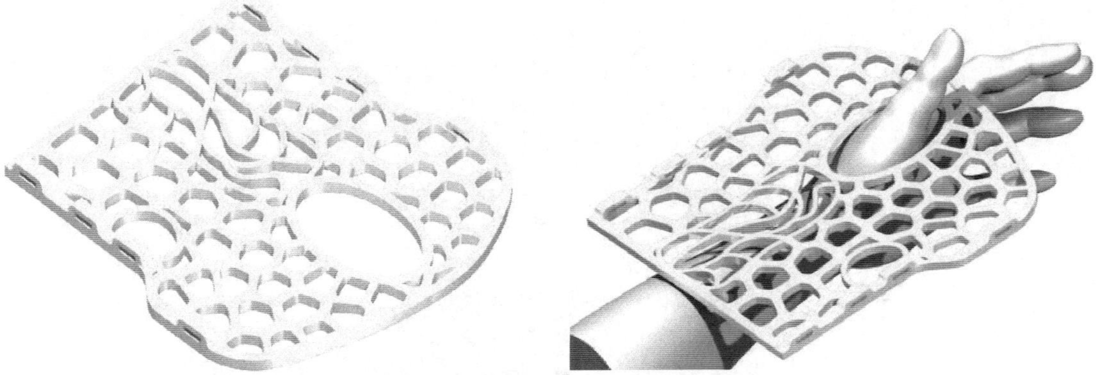

Figure 1. Thermoformable hand cast design concept.

2.2.1. Practical design. The cast sample was fabricated based on the average size of Malaysian adults' forearm and hand. According to Mohan *et al.* the average forearm and hand circumference for adults in Malaysia are 27.32 cm and 20.87 cm respectively, while the average palm length is 10.73 cm [11]. However, the cast can be fabricated into different sizes easily by changing the dimension scale in the 3D printing parameters setting. The maximum size of the cast that can be produced is only limited by the 3D printer printing area capacity; the bigger the machine's printing area capacity, the bigger the cast can be produced. The general dimension of the cast is illustrated in figure 2.

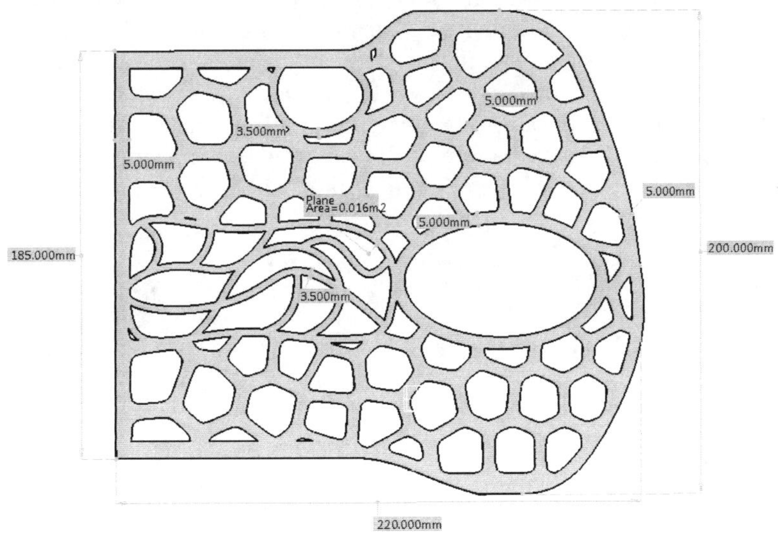

Figure 2. The thermoformable cast general dimensions.

The edge of the palm area was designed according to the palm's metacarpals position to enable the fingers to move freely. By unrestricting the fingers movement, users will still be able to utilize their hand for griping and holding while using the cast. A special pattern for the ventilation holes consisting of interconnecting curved lines was added along the forearm radius bone allowing the cast to be thermoformed easily according to the curvature between palm and wrist joint. The ventilated hole located at the end of the ulna's distal end or also known as ulnar head was produced larger and constructed with thinner round inner wall (3.5 mm). This feature provides a firm support structure to the ulna and also preventing skin irritation around the articulating joint. Figure 3 illustrates the design features of the thermoformable cast.

IOP Conf. Series: Materials Science and Engineering 1217 (2022) 012002 doi:10.1088/1757-899X/1217/1/012002

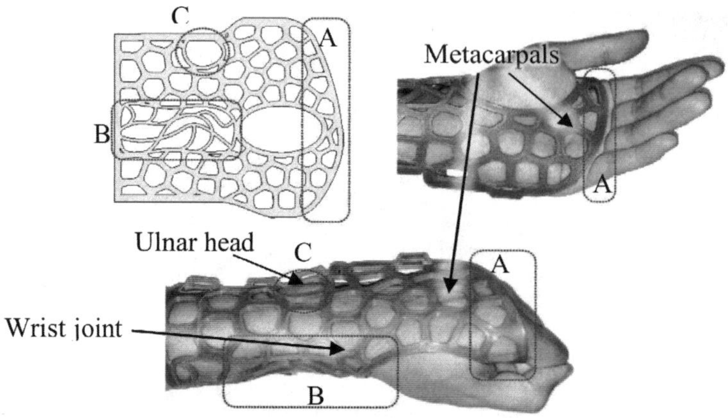

Figure 3. Design features of the thermoformable cast. (A) Unrestricted fingers movement; (B) Adaptive pattern for the wrist joint curvature; (C) Ulnar head opening.

2.2.2 Ventilation opening. The opening holes were arranged based on the honey comb lattice structure in order to ventilate the effected skin area effectively. The non-geometrical shapes of the holes provide more resistance of failure due to the accumulated stress in their vertices compared to the common geometrical figures [12]. With the ventilation holes, the total surface area of the cast has been reduced by 55%. The presence of the openings improves the hygiene of the affected area, avoid sweat accumulations, allergic problems, skin changes and allow a visual diagnosis by doctors.

The ventilation openings were developed with *sketch* tools in the *part modelling* workbench of CATIA V5 software. First, the *Voronoi* pattern of the cast was constructed with *spline* tool minimum 5mm apart from each other. The pattern then extruded into a solid part with 5mm of thickness (figure 4).

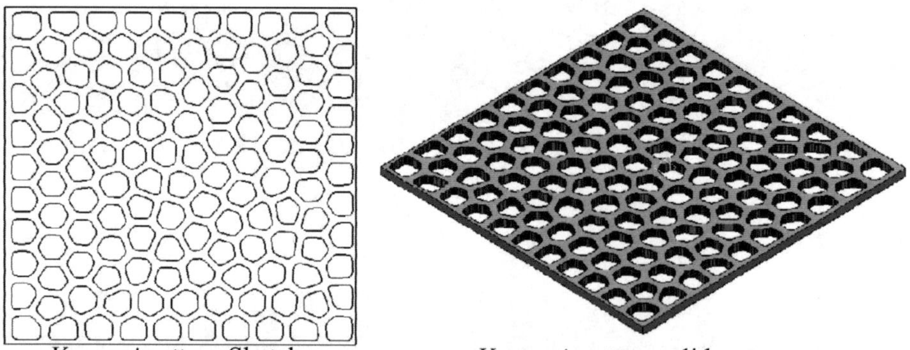

Voronoi pattern Sketch *Voronoi* pattern solid part

Figure 4. *Voronoi* pattern for the cast ventilation openings.

Next, a boundary defining the shape and size of the cast was constructed with *line* tool in the *surface modelling* workbench. The dimensions of the boundary were defined according to the average size of Malaysian adult's hand. The boundary line was then extruded as *surface* and used as *splitting object* to cut the *Voronoi* pattern solid part (figure 5).

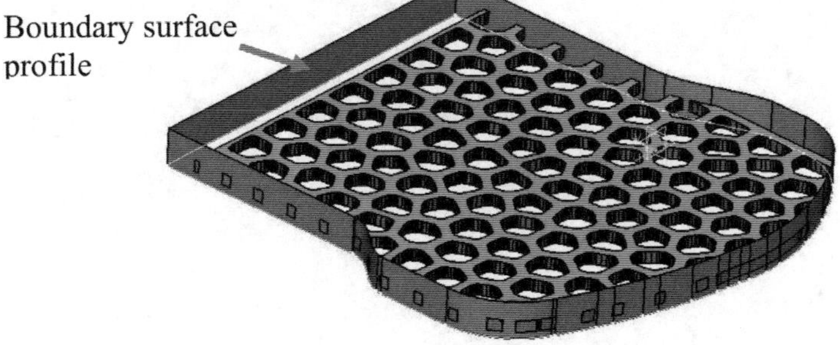

Figure 5. Cast boundary construction.

The thumb and the ulnar head openings position were determined based on a 3D-scanned hand model which was obtained from GrabCAD website. The model was re-scaled to the average hand size of Malaysian adults before being applied into the *surface modelling* workbench. Next, the surface area of the model that will be wrapped with the cast was identified and extracted with *extract* tool. After that, the edge of the cast design was aligned with the metacarpals position of the extracted surface by utilizing the software coordinate system. Once aligned, the opening profiles on the extracted surface were then projected onto the cast surface by using *transfer* tool. Finally, the openings for thumb and ulnar head were created with the projected profiles by using *sketch* and *pocket* tools in the *part modelling* workbench (figure 6).

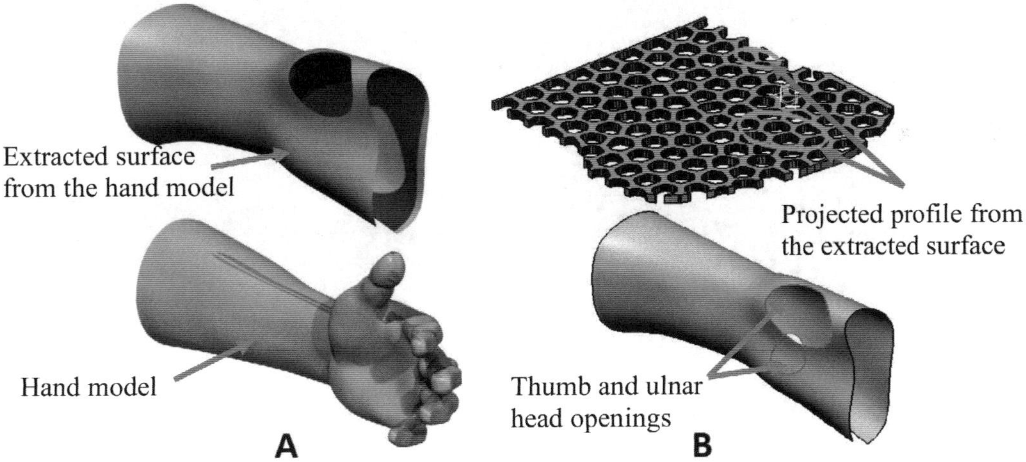

Figure 6. Thumb and ulnar head openings construction process. (A) Surface of the hand model that will be wrapped with the cast was extracted; (B) The opening profiles were transferred into the cast.

The interconnecting curve pattern along the radius bone was constructed with *sketch* and *extrude* tools in the *surface modelling* workbench. Once extruded, the surface profile was transferred into the *part modelling* workbench to convert the profile into a solid part by using *thickness* tool (figure 7).

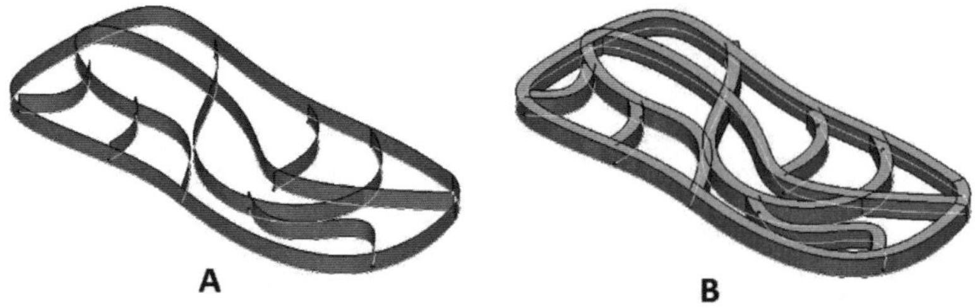

Figure 7. Construction of interconnecting curve pattern along the radius bone. (A) curved lines were extruded into a surface profile; (B) the surface profile was transformed into a solid part.

2.2.3 Cast thermoformability. The mechanical properties of the thermoformable cast were evaluated by using finite element analysis method with Autodesk Inventor Professional 2017 software. The flat design of the cast was first transformed into a hand-shaped cast by morphing it onto a hand model by using *wrap surface* tool in the *surface modelling* workbench. This method was also used as to mimic the plastic deformation of the 3D printed flat cast when it is heated and reshaped into a functional cast. Figure 8 illustrates the surface morphing process of the cast.

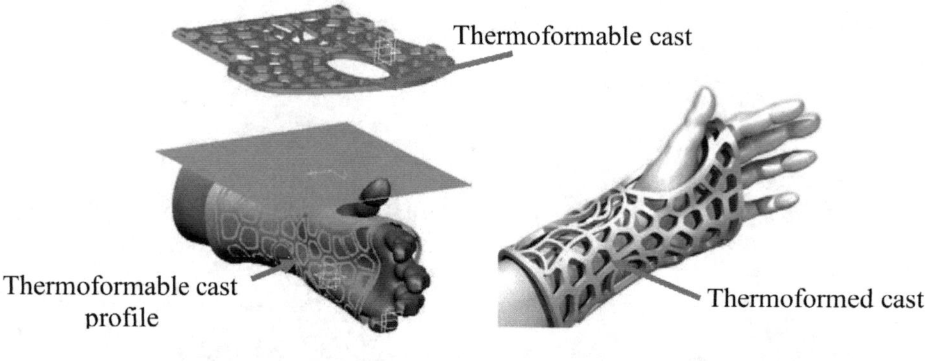

Figure 8. Cast surface morphing with Catia V5. The flat cast was morphed into a hand model(right) for Finite Element Analysis.

The cast thermoformability is the main aspect to determine how easily is it to be transformed and perfectly fitted to the user's arm. According to Kenning *et al.*, 3D printed models intended for thermoformed appliance must be printed with sufficient thickness to withstand thermoforming. Their study suggested that, to promote clinical utility, the minimum shell thickness of a 3D printed model for thermoformed appliances should be at least 2 mm. However, the thicker the shell, the harder the thermoforming process will be [13].

According to Bell, six to eight layers of plaster slabs and one to two layers for circumferential finger cast are usually sufficient [14]. Boyd *et al.*, recommended six to ten layers of plaster for the construction of adult's upper extremities orthotic cast [15]. A study by Pamar *et al.* stated that the thickness of eight layers of plaster cast which can withstand up to 366.67 N of loading is 6.21 mm. At that thickness, the plaster cast was heavy and bulky [16]. By considering the thermoformability factor, the loading requirements, and to reduce the bulkiness of the cast, the thermoformable cast design was set to 5 mm of thickness. The mechanical properties of the cast were then evaluated in order to determine whether the design could withstand the loading requirements or not.

2.2.4. *Constraint mechanism.* The thermoforming proses resulting a non-symmetrical shape of the cast making it difficult to be constraint onto the patient's arm with screws or other fixed fasteners. Moreover, inflammations and swellings that occur during the fracture treatments are common issues that must be accounted for during the splint design, as prescribed by Fitch *et al.* [17]. Hence, for this design, four plastic straps were employed for fastening the cast. This method also allows the cast to be rendered flexibly to accommodate limb swelling. Figure 9 illustrates the cast fastening method.

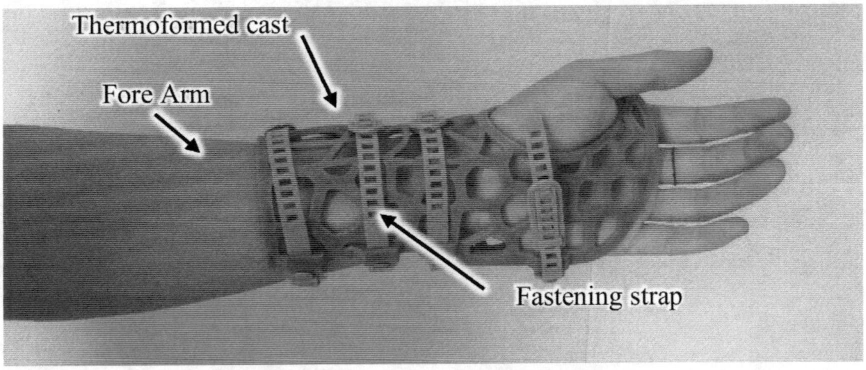

Figure 9. Cast fastening method.

2.3. *Mechanical loading*

A study regarding on the physical and mechanical properties of plaster of Paris bandage cast by Pamar *et al.* concluded that the load bearing capacity of the cast is proportional to its cross-section thickness. The maximum loading that can be withstood by an eight layers cast is 366.67 N [16]. In a previous study, Chen *et al.* have tested their 3D printed arm cast with 400 N of loadings and 1 N·m of bending moment along the palm movement directions [18]. Based on their findings, the proposed cast design is tested by exerting 400 N of loadings in X-axis and Z-axis directions, and 1 N·m of bending moment along the Y-axis direction of the cast. The directions of the loadings are based on the movement direction and the degree of freedom of a normal healthy hand as illustrated in figures 10 and 11. The base and the side surface of the cast opening are assumed to be fixed as in figure 12 as they are constrained by the plastic straps in real life application.

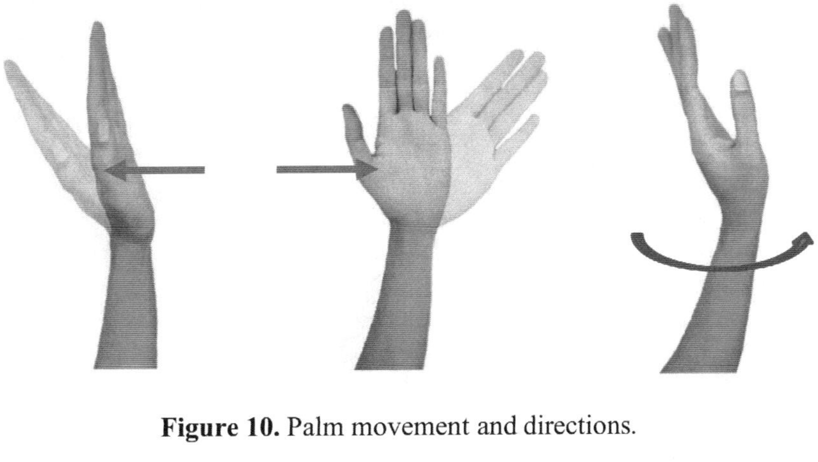

Figure 10. Palm movement and directions.

IOP Conf. Series: Materials Science and Engineering 1217 (2022) 012002 doi:10.1088/1757-899X/1217/1/012002

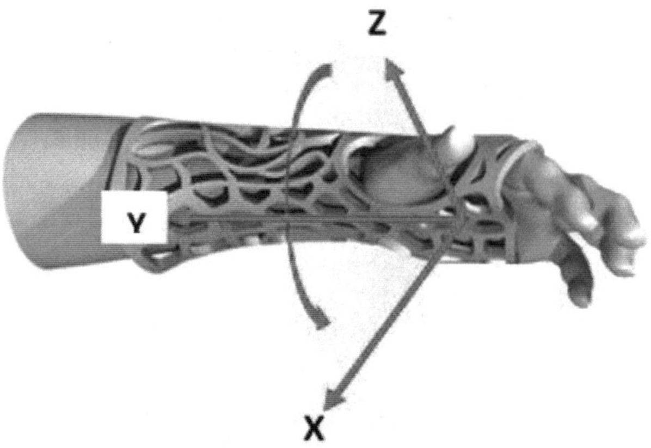

Figure 11. Definition of loading directions according to palm movement.

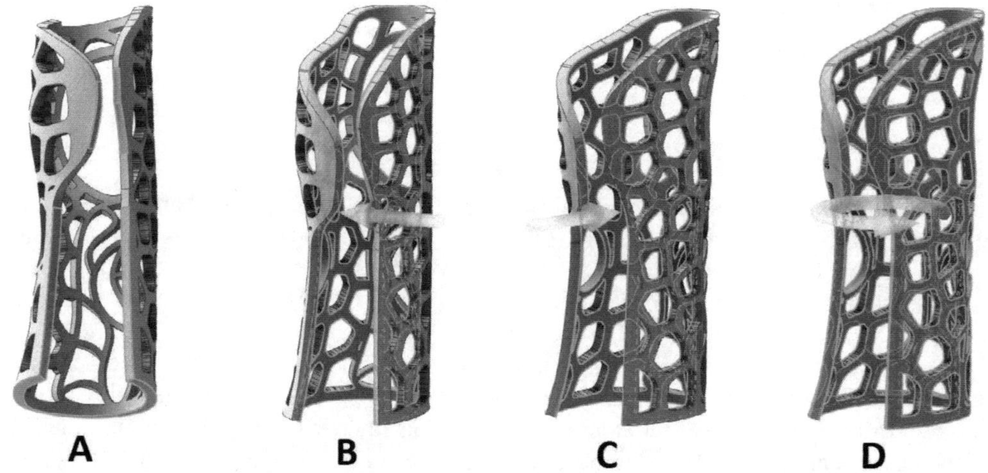

Figure 12. Constraint definitions and force distribution. (A) Constraint surface; (B) X-axis 400 N distributed force direction; (C) Z-axis 400 N distributed force direction; (D) 1 N·m bending moment along Y-axis direction.

2.4. Analysis parameters

Autodesk Inventor Professional offers Safety Factor (SF) calculations to be included in the analysis results. This feature enables user to predict the maximum loadings that can be exerted on the test object before it fails. In this study, the SF calculation was set based on the design's yield strength value. In general, yield strength is the stress a material can withstand without permanent deformation or a point at which it deforms plastically and no longer return to its original shape and dimension (by 0.2% in length). The calculation of SF based on the yield strength is often used to evaluate the loading limits of malleable materials.

Mesh structure was generated out of the cast model in order to identify the relationship between the stress concentrations and the affected region on the model due to loadings. The mesh sizing determines the accuracy of the program's calculations for the identification. The smaller the mesh size, the more complex the calculations and it cost the computer's memory consumptions. In this study, the mesh size on the complex faces and edges were defined much smaller than the average mesh size on the flat/noncomplex surface (as a fraction of average size = 0.2) to improve the simulations results. The mesh density distribution is illustrated in figure 13.

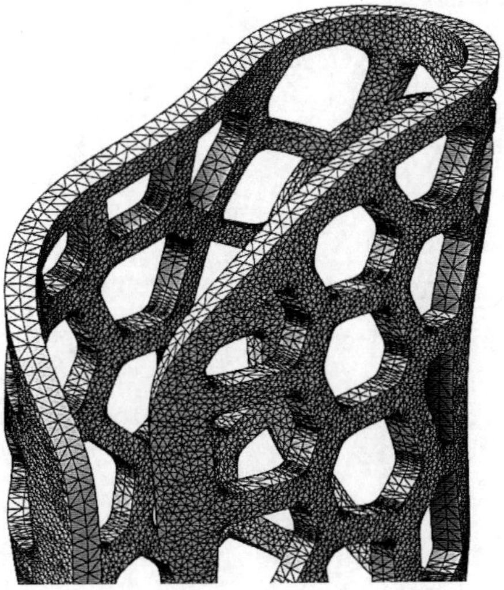

Figure 13. Mesh size definition and density distribution.

The mechanical properties of the material to be used for the fabrication of the model (PLA) are based on the data provided by the material's manufacturer as in table 1.

The distributed force location on the selected face of the model was placed on the centre of gravity of the model, which is calculated automatically by the program. The FEA parameters and setup for this design are described as in table 2.

Table 2. FEA parameter setup [10].

General objective and settings:	Study Type	Static Analysis
Mesh settings:	Avg. Element Size (fraction of model diameter)	0.1
	Min. Element Size (fraction of avg. size)	0.2
	Grading Factor	1.5
	Max. Turn Angle	60 deg
	Create Curved Mesh Elements	Yes
Material Mechanical Properties	Material	Polylactic Acid
	Yield Strength	49.5 MPa
	Ultimate Tensile Strength	45.6 MPa
	Young's Modulus	2.3465 GPa
	Poisson's Ratio	0.36 ul

2.5. Fabrication

The printing method for this cast is FDM. The flat thermoformable cast was printed with Ultimaker 2+3D printing machine and the STL file was sliced with Ultimaker Cura software. The printing parameters are demonstrated in table 3.

Table 3. Printing parameters.

Parameter	Setting
Material	PLA (Ultimaker)
Layer height	3 mm
Nozzle Diameter	6 mm
Infill	100 %
Support	Gyroid, Everywhere
Wall count	4
Top/Bottom Layers	3
Printing Temperature	200 ºC
Bed Temperature	60 ºC
Printing Speed	70 mm/s

3. Results

3.1. Von Mises Stress

Von Mises stress is a value used to determine whether the structure has started to deform plastically at any point. The Von Mises yield criterion states that if the Von Mises stress of a material under load is equal or greater than the yield limit of the material, then the material will deform plastically. Based on the simulation, the maximum Von Mises Stress value is 39.8572 MPa, which is resulting from the loading of the X-axis component (normal to the palm area). However, the Von Mises Stress maximum values for all loading conditions are lower than the yield limit of the material used. Table 4 and figure 14 summarised the analysis results.

Table 4. Von Mises Stress distribution of the cast under loadings.

Loading Component/Direction	Maximum Von Mises Stress (MPa)
X-axis	39.8572
Z-axis	30.1271
Bending Moment Along Y-Axis	0.03614

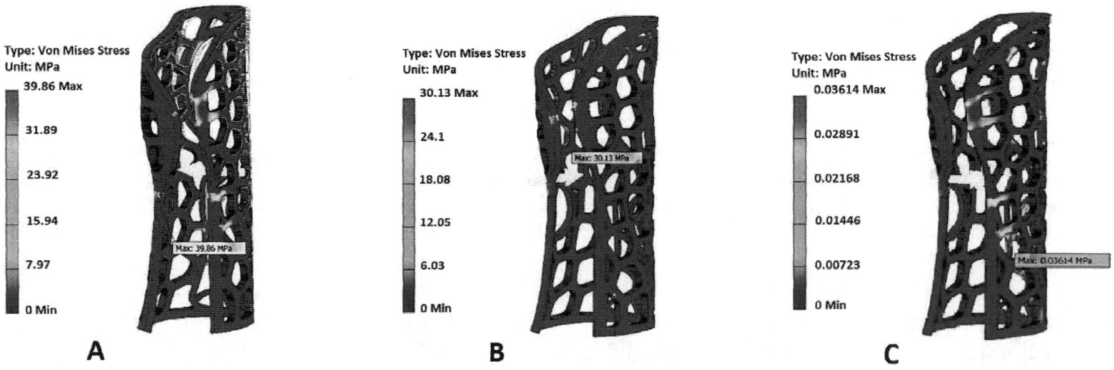

Figure 14. Von Mises Stress distribution of the cast under the load of 400 N and 1 N·m bending moment. (A) loading on X-axis direction; (B) loading on X-axis direction; (C) bending moment on X-axis direction.

3.2. Displacement

The maximum displacement obtained from the simulation is 3.29206 mm which is resulting from the X-axis loading. However, the deformation occurred is not a plastic deformation as the Von Mises Stress value does not exceed the material's yield limit. The design will be able to return to its initial shape when the loading is removed. The summary and illustration of the result is shown on table 5 and figure 15.

Table 5. Displacement of the cast under loadings.

Loading Component/Direction	Displacement (mm)
X-axis	3.292
Z-axis	0.747
Bending Moment Along Y-Axis	0.0003275

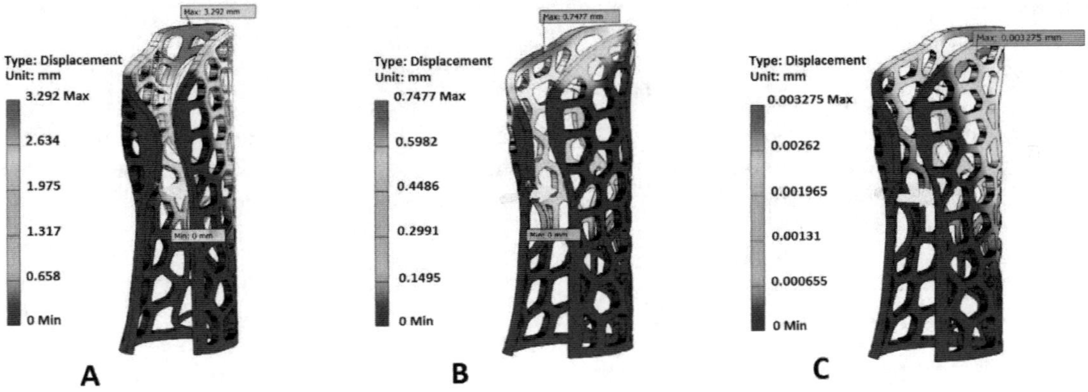

Figure 15. Displacement of the cast under the load of 400 N and 1 N·m bending moment. (A) loading on X-axis direction; (B) loading on X-axis direction; (C) bending moment on X-axis direction.

3.3. Safety factor

The calculation of safety factor for this cast is based on its yield strength value. The minimum value of 1.24 resulting from the X-axis loading explains that the design will not deform permanently under the given loading. The maximum value obtained from the simulation, which is 15, meaning that the elements on the certain region of the cast does not directly affected by the loading or the reaction force was too small and can be neglected. The summary and illustration of the results are shown in table 6 and figure 16.

Table 6. Safety Factor Value of the cast under loadings.

Loading Component/Direction	Minimum Safety Value
X-axis	1.24
Z-axis	1.64
Bending Moment Along Y-Axis	15

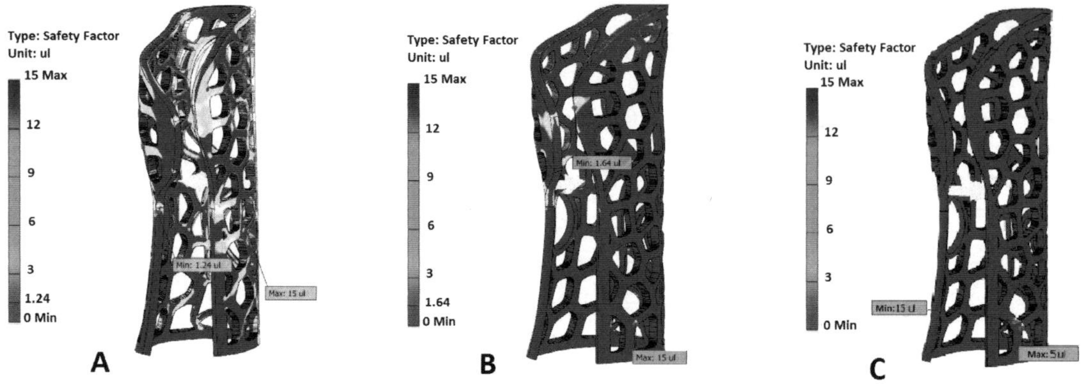

Figure 16. Safety factor value of the cast under the load of 400 N and 1 N·m bending moment. (A) loading on X-axis direction; (B) loading on X-axis direction; (C) bending moment on X-axis direction.

3.4 Fabrication
The 3D printing time for this cast with support structure is 3 hours and 15 minutes while the printing time without the support structure is 3 hours 14 minutes. The material consumption for both printing with support structure and without support structure shows a very small margin. The printing data and illustration are shown in the table 7 and figure 17.

Table 7. Thermoformable cast printing data.

Parameter	Value
Printing time (with support structure)	3 hours 15 minutes
Printing time (without support structure)	3 hours 14 minutes
Material consumption (with support structure)	13.27m/105g
Material consumption (without support structure)	13.23m/105g

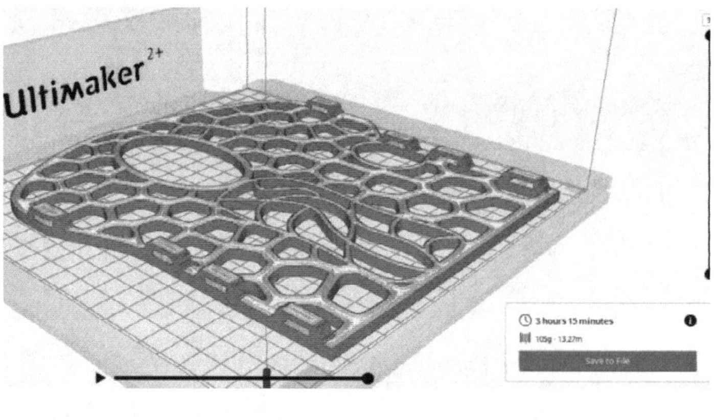

Figure 17. Thermoformable cast 3D printing.

4. Discussion

4.1. Data comparisons
Data obtained from this FEA was compared with data obtained from two previous study of Plaster of Paris mechanical properties by Parmar *et al*., and FEA results of 3D printed cast by SLS method by Chen *et al*. Parmar *et al*. investigated the mechanical properties of the plaster by mounting the sample

cast on universal testing machine (UTM - Autograph AG-100 KNG), and axial load was applied at 10 mm/min until failure [16]. However, Chen *et al.* does not mention the cast shell thickness used in their study. The data comparison is shown in table 8.

Table 8. Data Comparison of the Thermoformed 3D Printed Cast with Plaster of Paris Cast and SLS 3D Printed Cast.

Data/ Conditions	Plaster of Paris cast with 6.21 mm shell thickness [16]	3D Printed cast by Chen *et al.*[17]	Thermoformed 3D printed cast (5mm shell thickness)
Loading exerted (Condition)	366.67 N axial force (Limit before failure)	400 N axial force and 1 N·m bending moment (not mentioned)	400 N axial force and 1 N·m bending moment (Minimum Safety factor=1.24)
Global displacement (direction: normal to palm, perpendicular to palm, bending moment)	Break/Fail	6.168 mm, 3.123 mm, 0.226 mm.	3.292 mm, 0.747 mm, 3.275×10^{-4} mm
Global Stress (direction: normal to palm, perpendicular to palm, bending moment)	0.1 MPa	41.730 MPa, 22.067 MPa, 2.063 MPa	39.857 MPa, 30.127 MPa, 1.807×10^{-4} MPa

The maximum loading limit of the thermoformable cast before undergone plastic deformation is 496 N. Based on the comparison in table 8, the thermoformable 3D printed cast could withstand about 26% more loading capacity compared to the eight layers of plaster cast. Equation (1), (2) and (3) describe the calculation of strength margin between the Plaster of Paris and the Thermoformed 3D printed cast.

$$SF = \frac{F_t}{F_d} \tag{1}$$

The thermoformed 3D printed cast Safety Factor value is calculated as in equation (1) where, *SF* is the Safety Factor value of the thermoformed 3D printed cast obtained from FEA, F_t is the maximum loading limit that can be exerted on the thermoformed 3D printed cast before failure, and F_d is the designated loading which is 400 N of axial force that exerted on thermoformed 3D printed cast in X-axis direction. To determine the F_t value, equation (1) becomes:

$$\begin{aligned} F_t &= SF \times F_d \\ &= 400N \times 1.24 \\ &= 496N \end{aligned} \tag{2}$$

The strength margin, *SM* between the Plaster of Paris and the thermoformed 3D printed cast is determined in equation (3), where *SM* is the strength margin, F_t is the thermoformed 3D printed cast maximum loading obtained in equation (2), and F_p is the Plaster of Paris maximum loading which is 366.67N.

$$\begin{aligned} SM &= \frac{(F_t - F_p)}{F_t} \times 100\% \\ &= 26.075\% \end{aligned} \tag{3}$$

4.2. Strength analysis accuracy

The FEA simulation results may differ from the physical test made on the thermoformed cast. In the simulations, there were several considerations that were not taken into account such as:

a. The cast shell thickness distribution which is not completely uniform due to the thermoforming process might affecting the overall structural strength of the cast.
b. The mechanical properties of the printed parts might have differed from the data sheet provided by the manufacturer due to the printing process, parameters, and techniques. All the data from the data sheet were obtained from "ideal conditions" and may not reflect the real ones as we have printed.
c. The plastic straps used as fasteners and the arm bones supporting the cast were not included into the test.
d. The thermoformed cast mechanical properties might have changed due to the post-printing re-heating and cooling process.
e. The FEA simulation was based on the assumption that the model tested was not 3D printed. Thus, the 3D printing aspects such as the inter layer bonding and anisotropy effect were not taken into account.

4.3. Mechanical loadings

According to Chen *et al.*, the 400 N of mechanical loading used in the FEA is larger than the loading conditions encountered by healthy human in their daily life [17]. In reality, the mechanical loadings that will be endured by an injured human is far less compared to a healthy human as their activities and movement are limited. Due to the lack of data, the loadings applied in this study were based on suggestions by professionals and previous studies as mentioned before.

4.4. Load distribution and critical points

Based on the safety factor simulation, the lowest readings are mostly found at the sharp corners of the cast structure especially around the ventilation holes. These edges should be eliminated by using fillet features to improve the structure strength.

4.5. Cast 3D printing process

One of the advantages of this cast is, as a flat object, it does not require to be printed in high resolutions. Printing the cast in low resolutions shorten the printing time consumptions greatly. Theoretically, printing with higher layer thickness and larger nozzle size promotes better structural strength. As the nozzle movement are reduced, the time taken for the nozzle to deposit the molten filament layer by layer is also reduced; allowing the molten filament to be deposited on top of the initial layer more quickly before the initial layer cools off. A thicker layer height also prevents the layer to cool down quickly. These printing characteristics provide the cast a better inter layer bonding. The anisotropy structure of the cast was also minimized as the layers are extruded normal to the direction of the mechanical loadings.

5. Conclusion

The FEA simulation result shows that the thermoformable 3D printed PLA cast is stronger than the Plaster of Paris cast at the equivalent thickness. At lower production time compared to the existing 3D printed cast produced by normal printing method and conventional plaster cast, the thermoformable cast offers the same performance for its purpose. However, physical test on the thermoformed cast is advised in order to obtain more accurate and realistic data. Experimental work will be conducted in the future to validate the simulation result. Loading conditions for injured human need to be studied more and applied in the simulations so the cast design can be optimized accordingly.

References

[1] Fitzpatrick P A 2017 Design of a Patient Specific, 3D printed Arm Cast *KnE Engineering* **2** 135-42

[2] Chen Y-J, Lin H, Zhang X, Huang W, Shi L and Wang D 2017 Application of 3D–printed and patient-specific cast for the treatment of distal radius fractures: initial experience *3D Printing in Medicine* **3** 11

[3] Jiang J, Xu X and Stringer J 2018 Support Structures for Additive Manufacturing: A Review *J. of Manufacturing and Mat. Proc.* **2** 64

[4] Li J and Tanaka H 2018 Rapid customization system for 3D-printed splint using programmable modeling technique – a practical approach *3D Printing in Medicine* **4** 5

[5] Zhang D and Zhang X 2021 Rehabilitation Brace Based on the Internet of Things 3D Printing Technology in the Treatment and Repair of Joint Trauma *J. of Healthcare Eng.* 1-11

[6] Allum J, Moetazedian A, Gleadall A and Silberschmidt V 2020 Interlayer bonding has bulk-material strength in extrusion additive manufacturing: New understanding of anisotropy *Additive Manufacturing* **34** 101297

[7] Yang C and Hung S-W 2004 Modeling and Optimization of a Plastic Thermoforming Process *J. of Reinforced Plastics and Composites* **23** 109-121

[8] Ilyas R A, Sapuan S M, Harussani M M, Hakimi M Y A Y, Haziq M Z M, Atikah M S N, Asyraf M R M, Ishak M R, Razman M R, Nurazzi N M, Norrahim M N F, Abral H and Asrofi M 2021 Polylactic Acid (PLA) Biocomposite: Processing, Additive Manufacturing and Advanced Applications *Polymers* **13** 1326

[9] Jamshidian M, Tehrany E A, Imran M, Jacquot M and Desobry S 2010 Poly-Lactic Acid: Production, Applications, Nanocomposites, and Release Studies *Comprehensive Reviews in Food Science and Food Safety* **9** 552-71

[10] Ultimaker 2020 Ultimaker PLA SDS https://support.ultimaker.com/hc/en-us/articles/360012759359

[11] Mohan V, Shamsaimon N S, Japri M I, Yasin N E, Henry L J and Othman I R 2014 Fore arm circumference and hand length predict maximal hand grip strength among Malaysian *Middle-East J. of Sci. Res.* **21**(4) 634-639

[12] Blaya F, Pedro P S, Silva J L, D'Amato R, Heras E S and Juanes J A 2018 Design of an Orthopedic Product by Using Additive Manufacturing Technology: The Arm Splint *J. of Med. Systems* **42** 54

[13] Kenning K B, Risinger D C, English J D, Cozad B E, Harris L M, Ontiveros J C and Kasper F K 2021 Evaluation of the dimensional accuracy of thermoformed appliances taken from 3D printed models with varied shell thicknesses: An in vitro study *Int. Orthodontics* **19** 137–46

[14] Bell-Krotoski J A 2011 Tissue remodeling and contracture correction using serial plaster casting and orthotic positioning *Rehabilitation of The Hand and Upper Extremity* **6** 1599–609

[15] Boyd A S, Benjamin H J and Asplund C 2009 Splints and casts: indications and methods *Am Fam Physician* **80** 491

[16] Parmar A J, Tyagi S K, Dabas V S, Mistry J N, Jhala S K, Suthar D N, Bhatt R H, Pansuria D v. and Bhatti I M 2014 Assessment of the physical and mechanical properties of plaster of Paris bandage cast used as a splinting and casting materials *Veterinary World* **7** 1123-6

[17] Fitch M T, Nicks B A, Pariyadath M, HD M G and Manthey D E 2008 Basic splinting techniques. *N Engl J Med.* **359**(26) e32.

[18] Chen Y, Lin H, Yu Q, Zhang X, Wang D, Shi L, Huang W and Zhong S 2020 Application of 3D-Printed Orthopedic Cast for the Treatment of Forearm Fractures: Finite Element Analysis and Comparative Clinical Assessment *BioMed Res. Int.* 1-12

Mechanical properties of eggshells powder reinforced recycled high-density polyethylene

Y Y Farm[1,2*], A A A Mohammad Sappa[2], Z Mustafa[3], W K Muzammil[1,2], M A Ismail[1,2], N A Amaludin[1,2], A Kiring[2]

[1]Centre of Research in Energy and Advanced Materials, Faculty of Engineering, Universiti Malaysia Sabah, Jalan UMS, 88400 Kota Kinabalu, Sabah, Malaysia.
[2]Faculty of Engineering, Universiti Malaysia Sabah, Jalan UMS, 88400 Kota Kinabalu, Sabah, Malaysia.
[3]Faculty of Manufacturing Engineering, Universiti Teknikal Malaysia Melaka, Hang Tuah Jaya, 76100, Durian Tunggal, Melaka, Malaysia.

[*]Corresponding author: yanyan@ums.edu.my

Abstract. This study investigated the effect of thermal treated eggshells powder on the mechanical properties of recycled high-density polyethylene (r-HDPE) composite. Thermal treated eggshells powder, which was labelled as carbonised eggshells powder (CESP), was prepared at a low pyrolysis temperature of 400°C for one hour. The mechanical performance of the polymer composite reinforced with CESP was compared with the untreated dried raw eggshells powder (ESP) reinforced polymer composite. Each composite sample was prepared with a weight ratio of r-HDPE/filler varied at 90/10, 80/20, and 70/30. The polymer composite was prepared through a melt blending method in an internal mixture at 175°C at 32 RPM mixing rate. The composite samples were tested flexural analysis, while structure and morphology were characterised using Fourier Infrared Transformer (FTIR) and Scanning Electron Microscope (SEM). Based on the finding, the interfacial adhesion of the composite between filler and matrix has improved with ESP and CESP, respectively. The composite with CESP filler at 70/30 shows the optimum flexural strength and modulus of 32.21 MPa and 2.06 GPa, respectively. Overall, the introduction of CESP offers a better performance in improving the mechanical properties of the composite as compared to ESP.

1. Introduction

In a day, an average Malaysian household disposed of about at least 0.85 kg to 1.5 kg of plastic packaging per person, depending on their economic and geographical area [1]. Out of the total mass production of plastic packaging, only 24% of it being recycled [1]. In Malaysia, polyethylene terephthalate (PET), high-density polyethylene (HDPE) and polypropylene (PP) are the most common plastics that were recycled [2]. However, the limitation of recycled plastic, such as weak core strength value, has made the manufacturer lose interest in using recycled plastic as a matrix [3]. Thus, the composite material was introduced where one or more materials were combined into a new composite to achieve superior properties compared to their constituents [4].

Past studies related to the polymer composite had been successfully acknowledged improving its mechanical properties, including thermal stability, flexural and tensile strength, and cost reduction [5]. The filler occupies the space between the monomer of polyethylene and improves plastic moldability,

Content from this work may be used under the terms of the Creative Commons Attribution 3.0 licence. Any further distribution of this work must maintain attribution to the author(s) and the title of the work, journal citation and DOI.
Published under licence by IOP Publishing Ltd

heat-deflection, reduction of thermal expansion, and stability. Applying fillers such as calcium carbonate ($CaCO_3$), silica, and talc have improved polymer properties [3]. The dimensional stability of the composite was enhanced when the inorganic filler was introduced to the neat polymer matrix, with other properties such as modulus, heat distortion temperature, and even reduced the production cost [16].

The calcium carbonate particles have been used in thermoplastic manufacturing to reduce the amount of polymer and improve the mechanical properties of the polymers [3]. Hassan *et al.*, [17] claimed that the addition of filler introduced to the neat polymer has reduced the composite's elasticity. Eggshells contains up to 95% of $CaCO_3$ in the form of calcite. The eggshell powder is an inorganic material that requires chemical modification to enhance the interaction between the thermoplastic matrix and inorganic filler. This idea was supported by the study conducted by Senthil and Madan Raj [4] as the thermoplastic is hydrophobic; meanwhile, the eggshell powder is hydrophilic, and the chemical treatment or modification toward the eggshells powder created an interaction "bridge" between the filler and the thermoplastic matrix. The reinforcement of eggshells powder was intended to improve the load transfer when high force is applied toward the polymer composite especially when related to high duty applications such as piping or even furniture. For the industrial application, the polymer composite may not only be exposed to high load duty; but exposed temperature and humidity are also some of the concerning factors.

Thermoplastics are polymers that can be reprocessed through thermoforming, extrusion, or injection moulding with the application of heat [9]. Based on the performance of eggshells powder reinforcement in various thermoplastic such as LDPE [6], HDPE [7], and polypropylene (PP) [8], a new study was proposed based on the mechanical behaviour of eggshells powder reinforced high-density polyethylene [7]. The new study applied a smaller particle size at 32 μm and the chemical treatment for the inorganic material to enhance the interaction between the recycled matrix and the filler. In Siti Shuhadah *et al.* [6] study, the tensile strength of the LDPE- eggshell powder composite was reduced from 13 MPa to 7 MPa with the filler particle size of 63μm, where the eggshell powder loading was increased from 0 wt.% to 25 wt.%. The researchers explained this was due to poor adhesion between the inorganic filler and the particle agglomeration.

On the other hand, Buakaew *et al.* [7] studied the eggshells particle size range between 60 to 45 μm, 32 to 45 μm, and less than 20 μm reinforced with virgin high-density polyethylene (HDPE). In contrast, there was an improvement in tensile strength and flexural modulus from 2% to 13.9% for 5 wt.% to 20 wt.% filler content, respectively. When the filler loading was increased from 5 wt.% to 20 wt.%, the matrix and filler cooperation provided a stable load transfer "bridge" to overcome the applied load. The improvement in mechanical properties was evident where the particle size of the filler influenced the dispersion state of the reinforced polymers [18]. Aydemir [16] stated the mechanical properties of filled polymer emerged throughout the years, thus creating a broad foundation for the new study related to the mechanical properties of the reinforced polymer. However, the lack of finding using recycled matrix made room for this study and created the opportunity to discover the usage of the waste thermoplastic packaging in the hope to be applied for other applications.

Although the strength of the thermoplastic has a high possibility to be reduced during the manufacturing process of the prior purpose of thermoplastic [10], the recycled high-density polyethylene (r-HDPE) was chosen as the matrix. The lightweight of HDPE has a better performance than LDPE, and linear low-density polyethylene (LLDPE) has become the main reason for selecting HDPE as the polymer matrix. So far, there has been no study using eggshell-used HDPE to determine the mechanical behaviour of the polymer composite. The abundance of plastic packaging made up of HDPE and eggshells waste motivates to investigate the effect of thermal treated and untreated eggshells powder reinforced with r-HDPE. Thus, the outcome from this study which focused on the compatibility of eggshells as filler to r-HDPE, would create a new solution for the plastic waste and even put a value on the remanufactured waste plastic.

2. Materials and methods

2.1 Eggshells powder preparation

Eggshells were collected from the housing area in Alor Gajah, Melaka. The eggshells were washed, separated from the eggshell's membrane, and dried at room temperature for 24 hours. The eggshells were grounded using a dry blender and sieved to 32 μm using a sieve shaker (Octagon Sieve, model 200). The eggshells powder was dried overnight in the oven at 80°C and kept in a desiccator. The CESP was prepared by treating the eggshells powder with 85% phosphoric acid solution (R&M Chemical) at a weight ratio of 3:1. The mixture was mixed well and stored at room temperature. After 24 hours, the mixture was rinsed with distilled water until the filtered water turned colourless. The mixture was left dried overnight and carbonised using an electric furnace at 400°C for an hour.

2.2 Recycled HDPE composite preparation

The recycled HDPE samples were collected from used dishwasher bottles, detergent bottles, and shampoo bottles. All containers were washed, cleaned, and labels were removed to avoid any contamination. The containers were then cut to small pieces around 3 mm. The small pieces of r-HDPE were melted using an internal mixer (HAAKE PolyLab OS RheoDrive 16) at 175°C with 32 RPM. Eggshell powders with a specific ratio, as shown in table 1, were added into the melted r-HDPE. The mixing process continued for another 8 minutes to obtain constant torque, where the whole process took 10 minutes. The crusher was used to break down the extracted sample from the internal mixer into a smaller size for compression moulding (Geotech Hydraulic Hot Press). The compression moulding process involves three stages: preheating at 175°C for 3 minutes, followed by compression for 4 minutes at 15 tons. The quenching process took 3 hours without moving the specimen from the heating plate to avoid the formation of air bubbles. The specimen dimensions were 4 mm thickness, 64 mm support span with 6.4 mm overhanging, and 16 mm width.

Table 1. The composition of polymer composite.

Specimen	Matrix (wt.%)	Filler (wt.%)
r-HDPE	100	0
r-HDPE / 10 ESP	90	10
r-HDPE / 10 CESP	90	10
r-HDPE / 20 ESP	80	20
r-HDPE / 20 CESP	80	20
r-HDPE / 30 ESP	70	30
r-HDPE / 30 CESP	70	30

2.3 Infrared spectroscopy analysis

FTIR characterised the chemical bonds of ESP and CESP to compare the functional group presence between treated and untreated ESP. The analysis study within the wavenumber range of 4000-500 cm^{-1}.

2.4 Characterisation and mechanical testing

Flexural properties were determined according to ASTM D790-3 by using Universal Testing Machine (AGS-X Shimazu) with 1.77 mm/min crosshead speed with average of 5 tests for each composition. The specimens were conditioned at ambient temperature 23±2°C and 50±5% relative humidity. An average of five samples was used during the test. The neat r-HDPE, 30% ESP/ r-HDPE, and 30% CESP/r-HDPE specimens were used to study the morphology of the flexural fracture surface through Scanning Electron Microscope (Zeiss EVO 50). The specimen surface was coated beforehand with a thin layer of gold using a Quorum sputter coater.

3. Result and discussion

3.1 Eggshells powder preparation

The FTIR spectrum of ESP and CESP from the wavenumber range 500 cm^{-1} to 4000 cm^{-1} were presented in figure 1. The absorption band of each functional group was tabulated in table 2.

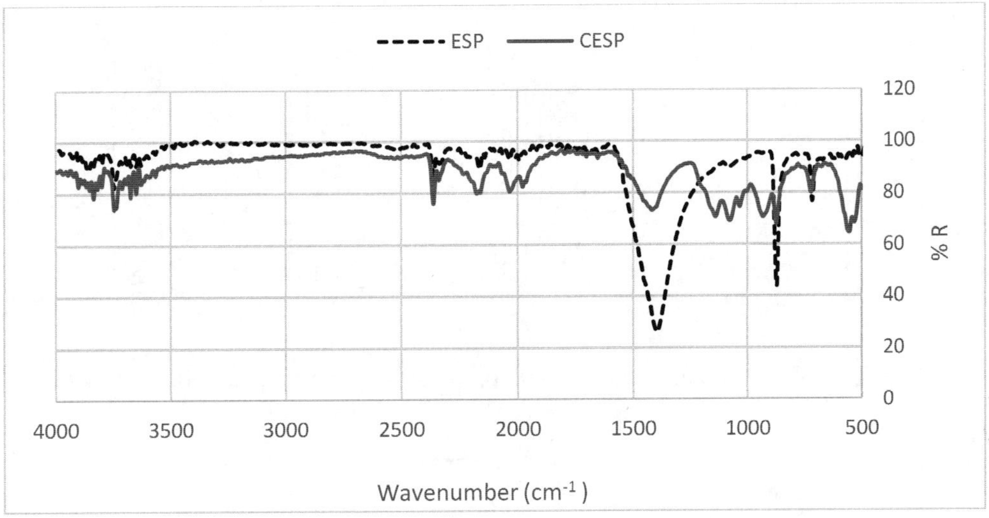

Figure 1. FTIR spectrum of dry eggshells powder.

Table 2. FTIR spectra of ESP and CESP.

Absorption bands	Wavenumber (cm^{-1})	
	ESP	CESP
OH^{-}	-	3875
CO_3^{2-}	1369	1417
CO_3^{2-}	871	873
CO_3^{2-}	711	715

A similar wave pattern was also obtained for other eggshells powder in earlier findings [11]. The difference in the absorption bands between two powders were due to the decreasing of reduced mass of functional groups connected with CO_3^{2-} molecules [11]. The calcination of CESP removed the protein strands, thus produced sharper bands for CO_3^{2-} molecules [12]. Based on figure 1, numbers of small bands can be observed in the FTIR spectrum as they represent the incomplete transformation of calcium oxide, CaO from calcite which may be resulted from the annealing temperature and duration. Thus, from the FTIR analysis, the energy band of CO_3^{2-} molecules increased as thermal treatment was done on the eggshells powder.

3.2 Mechanical properties

Figures 2 and 3 present the flexural strength and modulus of the composite at different filler loading, respectively. The bar chart pattern in figure 2 shows an increasing trend for flexural strength as the filler loading increased from 10 wt.% to 30 wt.%. A similar finding was reported by Siti Shuhada *et al.* [6] and Buakaew *et al.* [7]. Buakaew *et al.* [7] found that the maximum flexural strength and modulus for virgin HDPE/ESP were 31.2 MPa and 0.93 GPa, respectively at 30 wt.% of ESP. Thus, this study showed relatively higher flexural strength and modulus at 32.21 MPa and 2.06 GPa for r-HDPE/CESP at 40 wt.% filler content. According to Jacob *et al.* [11], filler reinforcement into the polymer composite exhibits high stiffness compared to the polymeric material. The stiffness of the polymer composite can

IOP Conf. Series: Materials Science and Engineering 1217 (2022) 012003 doi:10.1088/1757-899X/1217/1/012003

withstand a higher force. This observation resulted from the amalgamation of ESP particles into the composite, where ESP particles are stiffer than the r-HDPE.

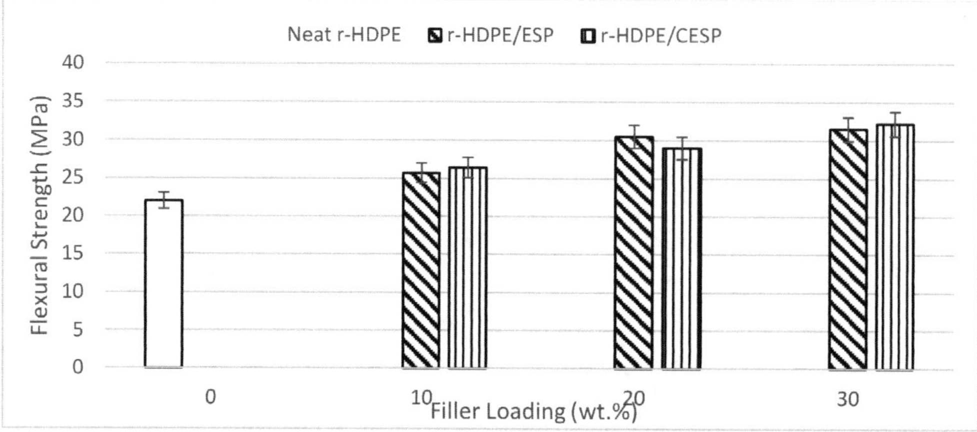

Figure 2. The flexural strength of composite at different filler loading.

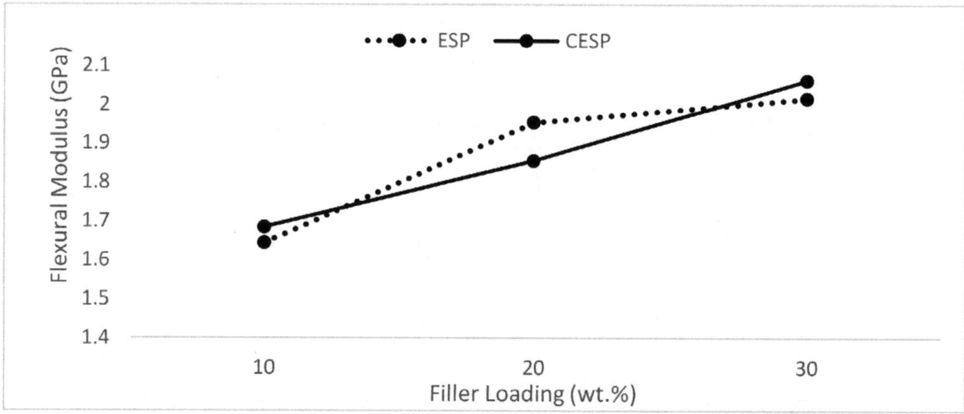

Figure 3. The flexural modulus of the composite at different filler loading.

Based on the data observation in figure 2, the flexural strength of CESP reinforced r-HDPE is higher than ESP reinforced r-HDPE. Siti Shuhadah *et al.* [6] stated that the bonding adhesion between the matrix and filler was considered weak without the chemical modification. This statement was supported by Hussein *et al.* [13] as the organic matter increases the hardness and stiffness of the particle when undergoing the carbonised process. The flexural modulus of composite shows the improved stiffness of composite with the addition of filler. A similar finding was also observed by Hanif *et al.* [14], where the resistance of polymer composite toward the elastic deformation increased with the increasing filler loading. The observation indicates the addition of filler for ESP was ideal at 20wt.%, whereas for CESP, the addition of filler improved the rigidity of composite till 30wt.%. The addition of filler at higher amounts resulted in an agglomeration that led to poor wetting and indirectly affected the stiffness of the composite. Although this study was using recycle matrix, the flexural strength and modulus of r-HDPE/filler were equivalent to the mechanical properties of HDPE/ESP composite, which used virgin HDPE as matrix and smaller particle size at 25 μm, which conducted by past researchers with maximum flexural strength recorded at 32.4 MPa with 40 wt.% filler loading [7]. This new formulated polymer composite can fit various manufacturing products, especially furniture such as cloth racks or even benches.

3.3 Morphology analysis

Figure 4 showed the composite microstructural image, which observed the formation fracture after the flexural test through Scanning Electron Microscope (SEM). Aside from the fracture within the specimen, the plastic deformation also can be observed based on the morphology analysis.

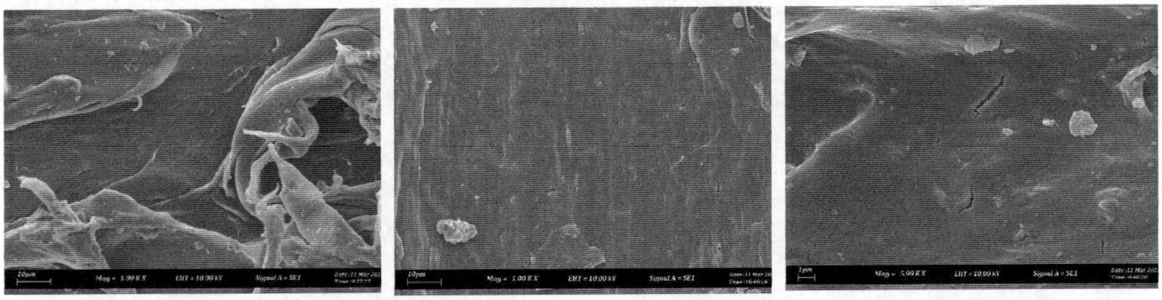

Figure 4. Microstructural image of neat r-HDPE (a), r-HDPE/30 wt.% ESP (b), and r-HDPE/30 wt.% CESP, respectively at *1*,000X magnification.

All three specimens experienced plastic deformation; however, the neat r-HDPE showed a rougher surface at 1,000 magnifications compared to reinforced r-HDPE. The addition of filler helps to reduce the plastic deformation, but poor interfacial bonding also can be observed between the matrix and filler. The advantage of using particle filler is less agglomeration of filler spotted through the morphology analysis. Azierah *et al.* [15] attributed the uniform distribution of filler into the composite to the particle size of eggshell particles.

4. Conclusion

The mechanical properties and morphological study of eggshells powder reinforced recycled high-density polyethylene were analysed. The filler dispersion into the matrix was proven capable of improving the strength of the composite. The reinforcement of CESP shows a better performance than ESP in enhancing the mechanical properties of composite material. The maximum flexural strength of the ESP and CESP in this study is 31.5 MPa and 32.2 MPa, respectively, at 30 wt.%. The morphological analysis also shows filler compatibility to interact with the matrix as only a small amount of filler agglomeration was observed. As the filler was reinforced into the composite, the fracture and plastic deformation was reduced, indicating the improvement in the strength of the polymer composite. The outlook for improved mechanical behaviour of any recycled plastic will be better in the future with the new accessibility to nanofiller and various chemical treatments to enhance the eggshells powder properties as filler in the aspect of geometry structure and chemical composition. The reinforcement of long fibre to the polymer also would improve the mechanical behaviour of polymer composite.

Acknowledgement

This work was supported by Universiti Malaysia Sabah (Grant number SGA0081-2019), Malaysia.

References

[1] Zainu ZA and Songip AR 2017 *J. Sci. Tech. Innov. Policy* **3** 18-22
[2] Chen HL, Nath TK, Chong S, Foo V, Gibbins C and Lechner AM 2021 *J. Springer Nature Appl. Sci.* **3** 437
[3] Lin Y and Chan CM 2012 *Adv. Poly. Nanocomposites* (Amsterdam: Woodhead Publishing) 55-90
[4] Senthil J and Madan P 2015 *Int. J. Mech. Eng. Robot.* **3** 7-18
[5] Ansari MNM, Mohammed L, Pua G, Jawaid M and Islam S. 2015 *Int. J. Poly. Sci.* **2015** 243947
[6] Siti Shuhadah M and Supri A G 2009 *J. Phys. Sci.*, **20** 87-98
[7] Buakaew W, Ruksakulpiwat Y, Suppakarn N, and Sutapun W 2013 *Adv. Mat. Res.* **699** 57-62

[8] Balogun O P, Adediran A A, Omotoyinbo J A, Alanme K K and Oladele I O 2020 *Int. J. Poly. Sci.* **2020** 8865748
[9] Mallick P K 2010 *Comp. Sci. Eng.* (Boca Raton: CRC Press) 174-207
[10] Adeniyi A, Agboola O, Sadiku E R, Durowoju M O, Olubambi P A, Babul R A, Ibrahim I D and Kupolati W K 2016 *Micro and Nano Technologies*, (United Kingdom: William Andrew) chapter 2 15-38
[11] Jacob M, Thonas S and Varughese KT 2004 *Comp. Sci. Tech.* **64** 955-65
[12] Wembabazi E, Mugisha P, Ratibu A, Wendiro D, Kyambadde J and Vuzi P 2015 *J. Spectrosc.* **2015** 714396
[13] Hussien A A, Salim R D and Sultan A A 2011 *J. Basrah Res.* **37** 36-42
[14] Hanif M P M, Jalilah A J and Naqiuddin M R M 2020 *IOP Conf. Ser.: Mater. Sci. Eng.* **864** 012122
[15] Azierah I, Wan Zarina W M and Azizah B 2018 *J. Poly. Sci. Tech.* **3** 29-34
[16] Aydemir B 2012 *J. Elastomers Plast.* **66** 41-44
[17] Hassan S B, Aigbodion V S and Patrick S N 2012 *Tribol. Ind.* **44** 217-25
[18] Igwe I O and Onuegbu G C 2012 *J. Poly. Sci.* **2** 56-61

IOP Conf. Series: Materials Science and Engineering 1217 (2022) 012004 doi:10.1088/1757-899X/1217/1/012004

Self-healing efficiency study of thermoset-thermoplastic polymer material

W H Choong*, H H Hamidi, K B Yeo

Mechanical Engineering Programme, Faculty of Engineering, Universiti Malaysia Sabah, Jalan UMS, 88400 Kota Kinabalu, Sabah, Malaysia.

*Corresponding author: whchoong@ums.edu.my

Abstract. This study is focused on exploring intrinsic self-healing polymer material development, where the inclusion of thermoplastic additives into thermoset polymer material as healing agents. Intrinsic self-healing thermoset-thermoplastic development is involving the material formulation of thermoset liquid resin (Poly Bisphenol A-co-epichlorohydrin) and thermoplastic (polycaprolactone). The material formulation ratio is up to 30% polycaprolactone with respect to thermoset weight. The mixture is heated and stirred to saturate at 80°C before the hardener is added. The mixture is cured and further finishing as Charpy impact test specimen. The specimen is fractured and absorbed impact energy property characterised through the Charpy impact test. The heat treatment is then performed to trigger the self-healing reaction in the polymer. The self-healing efficiency of the thermoset thermoplastic is investigated based on the absorbed impact energy before and after the heat treatment. The 20% or higher thermoplastic concentration in the polymer caused the polymer to possess high self-healing efficiency and faster healing time as compared to the low thermoplastic concentration polymer. However, the high concentration polymer has a disadvantage on the overall structural strength instead. On the contrary, 10% to 15% thermoplastic composition will result in lower and slower self-healing performance but higher initial structural strength.

1. Introduction

Despite the characteristics and physical properties of the polymer proved with tailor ability of high strength-weight ratio, the polymer's structure is very vulnerable to impairment in the form of crack. The impairment may find deep within the microstructure, where it is difficult to discover and almost impossible to mend [1]. Polymer material internal cracking will affect its functionality that is built for. It may further cause the polymer materials to expose to certain degree of mechanical degradation or failure. Internal cracks mainly lead by the micro cracking, which able compromise the structural integrity of the polymer and leads to disposal and wastage of the materials and resources.

Inspired by the self-healing of wounds in human skin, many researchers investigated on the method to develop the material with function of the self-heal ability to solve the cracking issue occurred in the polymer. Until now, three conceptual approaches that can exhibit the self-healing ability of the material: capsule based healing system, vascular healing system and intrinsic healing polymers [2]. The self-healing process can be either autonomic or non-autonomic, of which the autonomic process is the self-healing process that capable of conducted automatically without any human assistance, while the non-autonomic process is the process that requires external forces or pressure to trigger the self-healing process. Reversible bonding of the polymer is adapted from the famous Diels-Alder (DA) and retro-

Content from this work may be used under the terms of the Creative Commons Attribution 3.0 licence. Any further distribution of this work must maintain attribution to the author(s) and the title of the work, journal citation and DOI.
Published under licence by IOP Publishing Ltd

Diels Alder reactions (rDA) [2]. The self-healing ability of the polymer is gained by the reversible reaction of the polymer matrix. DA reaction itself allows the material to be able to react with the presence of the external source of energy and change from its monomeric state to cross-linked polymeric state or vice versa. The external source such as thermal to allow the cross-links bonds in the polymer to break down and then re-link upon cooling in healing the crack, demonstrated by Chen *et al.* (2002) applying above 120°C to allow the cross-links bonds in the polymer to break down and then re-link upon cooling [3].

Thermoset polymer materials (epoxy resin, polyurethanes, and synthetic rubber) are cured with cross-linked and form an irreversible chemical bond, where the cross-linking event limited the ability of remoulding or remelting. Self-healing or DA ability for thermoset material is crucial and it can be tailored by embedding with healing additives or agent such as thermoplastic. In focusing in developing the intrinsic self-healing thermoset material, the thermal diffusion healing agent should be based on reversibly bonded polymer material, limit effect on base material thermomechanical attributes and recuperation in strength, and minimum treatment temperature [4]. Some of the interested research works such as Luo *et al.* (2009), Meure *et al.* (2009), Ming and Min (2011), Nan (2013), Yufa and Sibener (2009) and Afl *et al.* (2019) are referred in developing the self-healing thermoset-thermoplastic polymer material [5 – 9]. The intrinsic self-healing thermoset-thermoset material self-healing efficiency is greatly on the material formulation, healing mechanism, treatment or triggering forces type or temperature of diffusion, treatment time, type of crack, repeatability, and others.

In this paper, the formulation and assessment of non-automic intrinsic healing polymer material is presented. The non-automic intrinsic healing polymer material is formulated based on the thermoset base material with thermoplastic material as the healing agent or additive material. The material is formulated in series of composition between the thermoset-thermoplastic material compositions for identifying the self-healing efficiency property through impact property assessment.

2. Material formulation and assessment

The thermoset and the thermoplastic are two very different materials, their unique properties make them harder to merge and form a new substance. In this project, the thermoset base material of bisphenol A resin and epiclororeohydin (industrial known as the Epikote 828 with curing agent of Epikure F205) and thermoplastic healing agent material of polycaprolactone monomer (CAPA 6500) are utilized in formulating the thermoset thermoplastic self-healing polymer material. Epikote 828 is an undiluted clear difunctional bisphenol A/epiclororeohydin derived liquid epoxy resin, a common thermoset material for industrial application such in building construction to transportation part material [10]. The material mechanical properties are tailor-ability with formulation with types of additives or curing materials. Meanwhile, the thermoplastic material polycaprolactone monomer (industrial known as the CAPA 6500) is either in bead or fine particle of high molecular weight linear polyester derives from caprolactone monomer [11]. The material is common applied as adhesive applications, as well as for rapid prototype applications. Both Epikote 828 and Capa 6500 material are very suitable and able to pair up and cross-link together to form the self-healable thermoset thermoplastic polymer [12]. Mixture of the thermoset and thermoplastic material resulting a unique properties and characteristics at the same time. The reason that caused the high compatibility between both ingredients is the structure chain inside the polymer chain.

2.1 Self-healing material formulation

The self-healing material formulation is based on the thermoset:thermoplastic composition. The thermoplastic polycaprolactone mechanical properties such as strength property is low compared to the thermoset material, therefore the composition of thermoplastic should be limited to low percentage. According to the polymer rule of mixture, newly formulated material properties will tend to the higher composition material applicant. There are three different samples with the different composition of the material used and listed in table 1 below. Thermoset composition weight is included the thermoset curing agent weight, where the 50:50 (thermoset base material:curing agent) mixing ratio.

IOP Conf. Series: Materials Science and Engineering 1217 (2022) 012004 doi:10.1088/1757-899X/1217/1/012004

Table 1. Samples composition ratio.

Samples	Thermoset (%) : Thermoplastic (%)
1	70 : 30
2	80 : 20
3	85 : 15
4	90 : 10

2.2 Specimen fabrication

The material self-healing efficiency is assessed through the material impact property, where Charpy impact test is adopted. Therefore, the self-healing material specimen geometry is sized according to ISO 179 [13].

The mixture of thermoset and thermoplastic is prepared in a metal cup and heated at 80°C (above the polycaprolactone melting temperature which is between 58°C to 60°C) with a hot plate. The solution is continuous stirred till saturated. The thermoset curing agent is added and stirred till saturated. The curing agent will initiate the curing process. Saturated solution is poured into a preheated mould and cure at temperature control environment (60°C for 4 hours and rest till reach room temperature). Cured specimens with thickness of 4 mm are further processed to adequate dimension of 80 mm (length) x 10 mm (width) and 45° mid notch of 2mm depth.

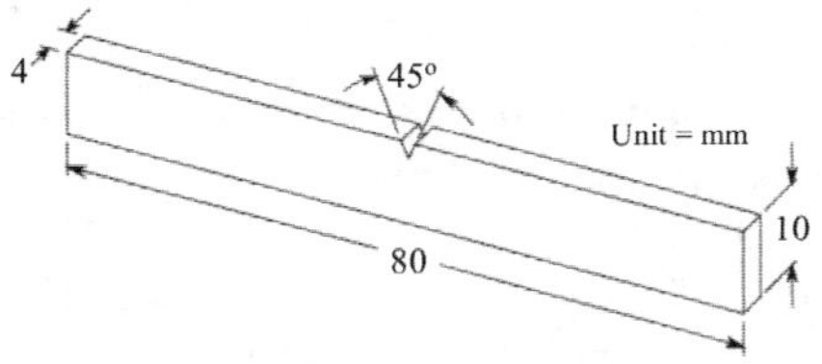

Figure 1. Charpy test specimen geometry detail.

2.3 Material self-healing efficiency assessment

The material self-healing efficiency is assessed through the material impact property, where Charpy impact test is adopted. Charpy impact test is assessing the material amount of absorbed energy before it fractured. Therefore, the self-healing material specimen geometry is sized according to ISO 179, shown in figure 1, and average weight of 3.43 gm. The material self-healing efficiency can be defined as:

$$\text{Self-healing efficiency,} \quad R = \frac{E'}{E} \times 100\% \quad (1)$$

where E and E' is the initial and treated specimen amount of absorbed energy, respectively. Specimens are tested and amount of absorbed energy is recorded. Then, the fractured specimens are mount into a treatment jig to ensure the two broken pieces (fractured specimen) is forming the initial shape and stay intact. The self-healing efficiency of the polymer is evaluated based on the capability of the polymer to heal the crack after the impact test and the structural strength of the polymer at the mended region. There are two treatment variables in determining material self-healing efficiency:

i. Treatment temperature; and
ii. Treatment time.

Figure 2 shown the treatment jig that used to hold the fractured specimen in place. The fractured specimen is pressed axially by a 0.1 Nm tightening screw torque. The 0.1 Nm tightening torque is an axial loading that just sufficient to hold the fractured pieces intact and will not deform it. For the first treatment trial, the treatment is conducted with set of treatment temperature (60, 70, 80 and 100°C) for 30 minutes. The second treatment trial, the treatment conducted with set of treatment period (10, 20, 30 and 40 minutes) with treatment temperature of 80°C. Treated specimen is undergo the Charpy impact assessment to obtain the impact energy property for the evaluating the material self-healing efficiency.

Figure 2. Treatment jig with holding the fractured specimen.

3. Result and discussion
In this study, the results and its discussion are elaborated as following:

3.1 Specimen fracture region before and after treatment
The specimen before and after treatment is shown in figure 3. Before the treatment is conducted, the specimen is fractured into two or three pieces. Upon the treatment process, the specimen facture pieces are bonded, and formed as single piece or initial state geometry form. For sample with thermoplastic composition of 20% and 30%, fracture lines are not visible and fully bonded or healed compared with lower thermoplastic composition, respectively. Thermoplastic property of remould-ability or spontaneously reform of intermolecular interaction embedded into the thermoset-thermoplastic sample.

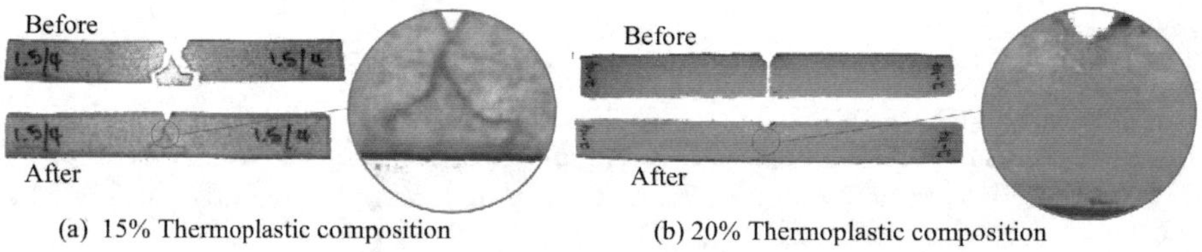

Figure 3. Specimen fracture region before and after treatment process (a) 15% thermoplastic composition and (b) 20% thermoplastic composition.

3.2 Self-healing efficiency assessment
The thermoset material is considered as a brittle material with the impact strength of 1.63 KJ/m^2 compared to thermoplastic material impact strength of 11.4 KJ/m^2. Based on the Charpy impact test, the thermoset-thermoplastic formulated material properties are greatly influenced by the thermoplastic composition, which the impact strength property of each composition applicants shown in figure 4. For the 10% and 15% thermoplastic composition, the material impact strength property is degraded about 29.5% and 31.9% compared to the pure thermoset material property, respectively. The higher thermoplastic composition percentage, the lower the impact strength or the material brittleness increase.

Impact strength property of each thermoplastic composition material applicants under before and after treatment condition is plotted in figure 4. The material applicants are treated under 80°C for 30 minutes. Upon the treatment, the applicants of high thermoplastic composition (20% and 30%) able to recover about 97% of its impact strength compared to lower thermoplastic composition (10% and 15%), which oleh recover about 38.3% and 28%, respectively. The impact strength recovery rate is high with higher thermoplastic composition, where the thermoplastic as the self-healing additive to the thermoset based material is activated.

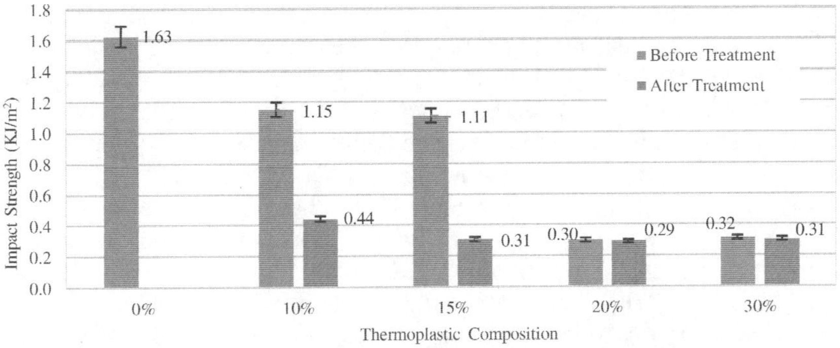

Figure 4. Average impact strength property of each thermoplastic composition materials with respect to before and after treatment condition (average coefficient of variance at 4%).

The study also assessed the treatment condition variables (temperature and time) in influencing the material self-healing efficiency. The materials self-healing efficiency respect to the treatment conditions is plotted in figure 5 below, and it generally shown that the self-healing efficiency is affected by the treatment temperature and amount of time. Treatment temperature of 70°C is considered the minimum treatment temperature with 30 minutes treatment for material applicants of above 20% of thermoplastic composition for achieving more than 80% self-healing efficiency. Meanwhile for the lower thermoplastic composition (below 20% composition), the self-healing ability or re-crosslinked event triggered upon the treatment under 80°C. The material applicant's self-efficiency is very low below 30%, even while the temperature increased up to 100°C. The high thermoplastic composition materials (20% and 30% thermoplastic composition) self-healing ability (self-healing efficiency of 40%) can be activated with treatment temperature of 80°C for 5 minutes treatment time. The material applicants self-healing efficiency can reach up to 92%, where the treatment time just prolong till 10 minutes and above. For lower thermoplastic composition (10% and 15%), the self-efficiency can only be achieved up to 40% as the treatment time prolonged. It also shown that the self-healing ability only be activated by the treatment time by 20 minutes for 15% thermoplastic composition and 30 minutes for 10% thermoplastic composition.

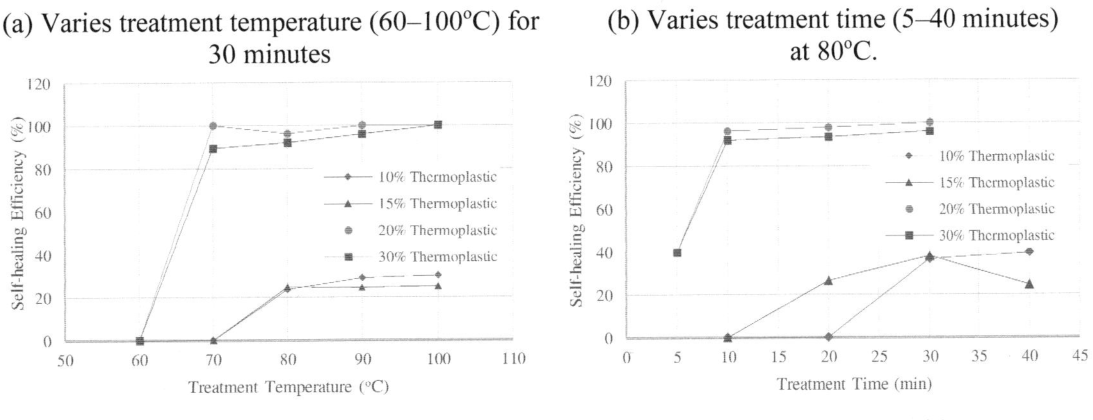

Figure 5. Self-healing efficiency of each thermoplastic composition material with respect to treatment temperature and time.

4. Conclusion

The study on self-healing efficiency with formulated polymer material composition of Epikote 828 (thermoset) and CAPA 6500 (thermoplastic). The formulated polymer material demonstrated the intrinsic self-healing mechanism, where applying heat source in triggering the healing ability. Charpy impact test utilized in determining the material self-healing efficiency. With higher composition of thermoplastic inside the polymer, the specimen will have greater self-healing performance than the low thermoplastic composition polymer. In exchange, the polymer is more brittle and low resistance or susceptible to impact cause fracture. For the low thermoplastic polymer, the polymer is tougher, the overall structural strength of the polymer is better than the high thermoplastic composition polymer.

The self-healing reaction in the thermoset thermoplastic polymer requires external stimulus to trigger it, in this case it is heat. The heat treatment temperature has a direct impact to the self-healing performance of the polymer, whether it is low or high thermoplastic composition polymer. The high thermoplastic composition polymer requires lower treatment temperature, the temperature around 60°C or 70°C will trigger the self-healing reaction while the low thermoplastic polymer requires temperature around 80°C. Furthermore, the treatment time is also a factor that affects the self-healing performance of the thermoset thermoplastic polymer. The low thermoplastic composition polymer requires at least 30 to 40 minutes to complete the self-healing reaction while the high thermoplastic composition polymer need only 5 or 10 minutes to allow the polymer to heal. It is safe to say that the effect of the composition ratio, treatment temperature and treatment time on the self-healing efficiency has been successfully investigated. The overall structural strength of the thermoset thermoplastic polymer is still very weak, although the self-healing performance of the polymer can compensate the structural strength problem, but the self-healable polymer is still not very practical in long term usage. Hence, it is important to perform more studies to understand the full capabilities of the self-healable thermoset thermoplastic polymer.

Acknowledgements

This study had been carried out by the researchers of Universiti Malaysia Sabah and funded by Ministry of Higher Education (ERGS0034-TK-1-2013).

References

[1] White S, Sottos N and Geubelle P 2000 *Nature* **409** 794
[2] Blaiszik B, Kramer S, Olugebefola S C, Moore J S, Sottos N R and White S R 2010 *Anual Reviews of Materials Res.* **40** 179
[3] Stefania D I, Alfonso M and Eugenio A 2018 *Paint and Coatings Industry* ed Faris Yilmaz, (London: Intechopen)
[4] Chen X., Dam M, Ono K, Mal A, Shen H, Nutt S, Sheran K and Wudl F 2002 *Science* **295** 5560
[5] Hayes S A, Zhang W, Branthwaite M and Jones F R 2007 *J. of The Royal Soc. Interface* **4** 381
[6] Luo X, Ou R., Eberly D E, Singhal A, Viratyaporn W and Mather P T 2009 *ACS Applied Materials & Interfaces* **1** 612
[7] Meure S, Wu D Y and Furman S A 2009 *Acta Materialia* **57** 4312
[8] Ming Q Z and Min Z R 2011 *Self-healing polymer and polymer composites* (London: Wiley)
[9] Nan B, Kei S and George P S 2013 *Polymer Chem.* **4** 724
[10] Afl M, Rahmathullah M and Palmese G R 2009 *J. of Applied Polymer Sci.* **113** 2191
[11] Hexion 2007 *Epikote resin 828 – Technical data sheet*
[12] Tri-Iso 2010 *Capa 6500 – Product data sheet*
[13] Cohades A, Branfoot C, Rae S and Bond I P 2018 *Advaced Mat. Interfaces* **5**
[14] ISO 179-1 2010 *Plastics – Determination of charpy impact properties*

Plastic straining and concomitant microstructure recrystallization of Ni-Cu alloy in the undercooled condition

H An[1,2,3*], N J Siambun[1], B L Chua[1], M J H Gan[1]

[1]Mechanical Engineering, Faculty of Engineering, Universiti Malaysia Sabah, 88400 Kota Kinabalu, Sabah, Malaysia.
[2]College of Mechanical and Electronic Engineering, Huanghe Jiaotong, University Jiaozuo People's Republic of China.
[3]Henan Engineering Technology Research Center on Intelligent Manufacturing Technology and Equipment, Jiaozuo, People's Republic of China.

[*]Corresponding author: anhongen@sina.com

Abstract. Microstructure and microtexture of rapidly solidified undercooled Ni-Cu alloys were investigated. The characteristic undercooling of Ni80Cu20 alloy was determined as 45K, 90K and 160K. Dendrite deformation due to rapid solidification led to strong deformation microtexture. Due to recrystallization upon annealing after recalescence, many subgrains were formed in the microstructure. Further, annealing the quenched alloy at 900°C, new microtextures and subgrains were formed, which was due to recrystallization and dislocation network rearrangement. The results of comparative experiment proved the recrystallization mechanism of the microstructure refinement in the non-equilibrium solidification structure of the undercooled binary alloy.

Keywords: Undercooling; Stress; Deformation; Microtexture; Subgrain.

1. Introduction

Grain refinement mechanism of undercooled solidification microstructure of binary alloys is a topic of debate for decades [1-16]. For highly undercooled single phase binary alloys, recent studies on Fe-Ni [3], Ni-Cu [4], Fe-Co [6] and Ni-C [9] alloys show that two grain refinement could be found. Microstructure refinement occurring in the small undercooling range is due to dendrite remelting [2, 5]. While the mechanisms for the grain refinement occurring in the large undercooling are proposed as follows: (a) the dendrite break-up caused by dendrite remelting [2] or strain effect upon recalescence ; (b) the copious nucleation [2] due to pressure pulse generated from the shrinkage cavity collapse [2]; (c) recrystallization during post-recalescence period [2]. etc. For Ni-Cu alloys, a widely used theoretical model for explaining the grain refinement event beyond the critical undercooling is Karma's dendrite remelting model [7]. Karma's model suggested that if the thermal plateau time (Δt_{pl}) during solidification is larger than the time for dendrite break-up (Δt_{bu}) by Rayleigh instability, the grain refinement due to dendrites remelting will occur. In addition, recrystallization mechanism seems also applicable to explain the grain refinement beyond the critical undercooling. However, the grain

Content from this work may be used under the terms of the Creative Commons Attribution 3.0 licence. Any further distribution of this work must maintain attribution to the author(s) and the title of the work, journal citation and DOI.
Published under licence by IOP Publishing Ltd

refinement mechanism of Ni-Cu alloys under large undercooling is controversial as listed above, and there is a lack of reliable experimental evidence.

In this paper, the microstructure evolution of experimental samples was studied. And the orientation information and the direct relationship between orientation and microstructure were obtained by further study of microstructure. Finally, it is confirmed that recrystallization leads to grain refinement of Ni-Cu alloy with large undercooling.

2. Materials and methods

The pure copper (99.99%) and pure nickel (99.99%) weighed in a certain proportion are smelted in-situ in a vacuum arc melting furnace under an argon atmosphere to prepare Ni80Cu20 master alloy. The alloy was remelted many times to ensure well-mixed, and 4g sample was cut from each original ingot for experiments. Quartz tube is put into the high frequency induction coil, and the vacuum of the experimental furnace is pumped to 3×10^{-3} Pa, argon backfilled to 5×10^{-2} Pa. An infrared thermometer was used to record the temperature curve during the solidification. Firstly, the temperature was slowly raised to 750°C for 15 minutes to melt B_2O_3 and fully cover the alloy. Secondly, when the temperature is higher than the melting point of the alloy from 100K to 200K and kept for 15 minutes, B_2O_3 purifier can fully absorb impurities to inhibit heterogeneous nucleation. And finally, the cyclic superheating method is used until the ideal degree of undercooling is obtained. The experimental diagram is shown in figure 1. After cutting, inlaying and polishing, the cooled samples were etched with 20 ml HCl and 20 ml HNO_3 mixed acid solution. Then, the microstructure characteristics of undercooled alloy were observed by metallographic microscope, and metallographic photos were taken.

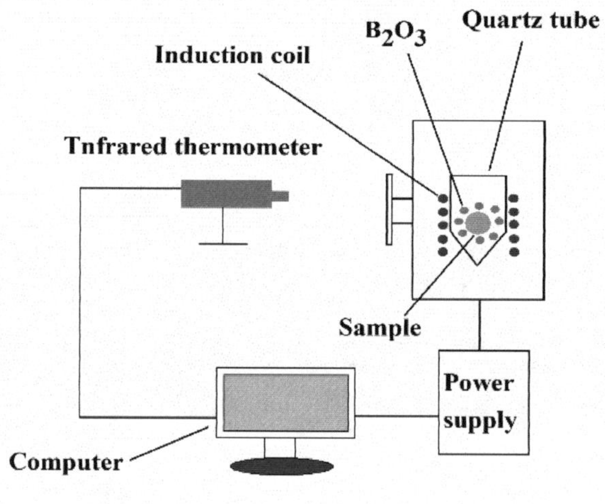

Figure 1. Schematic diagram of experimental device.

3. Results and discussions

As increasing undercooling, two grain refinements occur: one occurs at low undercooling, and the other occurs at large undercooling. The microstructure at different undercooling is shown in figure 2. In a small undercooling range, e.g. at an undercooling $\Delta T<\Delta T1=45K$, highly developed dendrites prevails (figure 2a). As increasing undercooling, dendrites gradually refined into smaller equiaxed grains due to dendrite remelting. At an undercooling range $\Delta T1$-$\Delta T2=90K$, refined grains are present in the solidification microstructure (figure 2b). A further increasing the undercooling leads to again coarse grains. As the undercooling $\Delta T>\Delta T3=110K$, the dominating dendrite growth mechanism has changed from solute diffusion to thermal diffusion and thus finer dendrite beams appear. At $\Delta T3$, few dendrite fragments can be found in the as formed microstructure, and only coarse dendrites can be observed (figure 2c). At an undercooling higher than a critical value ΔT^* (160K), grain refined microstructure

IOP Conf. Series: Materials Science and Engineering 1217 (2022) 012005 doi:10.1088/1757-899X/1217/1/012005

prevails again (figure 2d). The microstructure is fine equiaxed grains and some twins due to annealing indicate the occurrence of recrystallization.

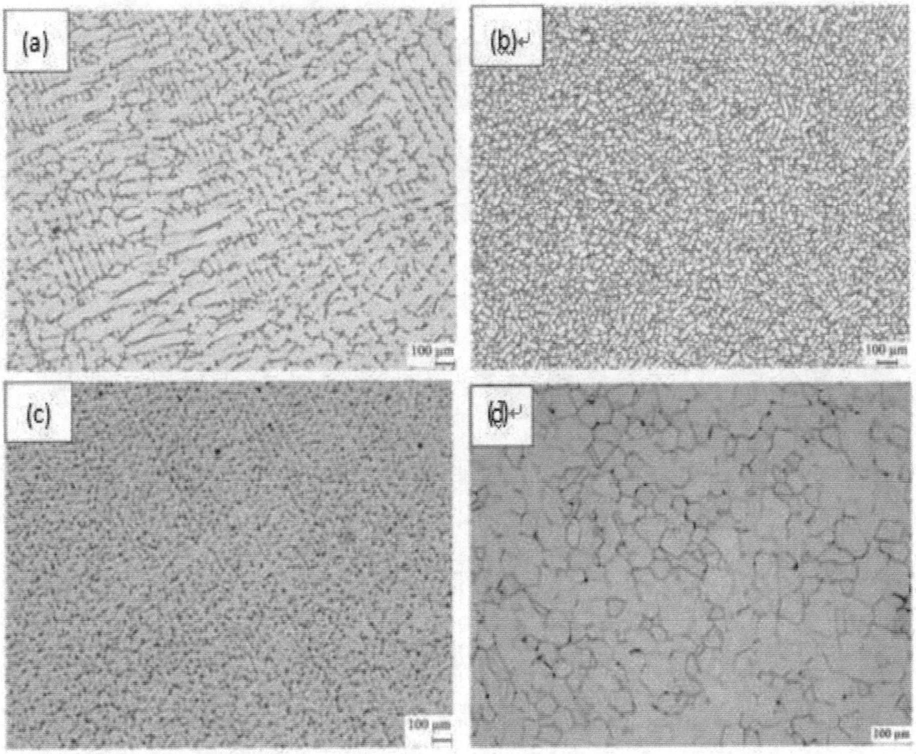

Figure 2. Microstructures at small undercoolings (a.18K; b. 70K; d. 155K; d. 165K).

If the undercooling increases to 300K, many small equiaxed subgrains with small angle grain boundaries (GBs) can be observed in the samples quenched before recalescence (figures 3a and 3b). In figure 3a, a large number of equiaxed subgrains were observed in the microstructure. Some GBs with medium axial subgrain in microstructure have large angle properties. figures 3d and 3e show grain boundary orientation and orientation distribution, respectively. Compared with most of the large angle grain boundaries in the naturally cooled alloy, nearly 71% of the small angle grain boundaries were found (figures 3b and 3c), and a high orientation texture was observed in the quenched alloy (figure 3d), in which the whole region has its own orientation, because the subgrains in the microstructure have almost the same orientation.

The quenched alloy was annealed at 1173K (900°C) for 30 minutes, recrystallization occurred, and more visible subgrain boundaries were observed (figure 4a-b). The grains were completely refined and only fine equiaxed subgrains were observed (figure 4a). It can also be observed that the fraction of small angle GBS increases to about 86% (figure 4d), while the fraction of large angle boundary is still very low (figure 4d). The grain boundary dislocation and dislocation distribution are shown in figure 4e. Compared with figure 3e, the scores of large angle GBS and Σ3 boundaries in figure 4e are reduced to 14% and 0.2%, respectively. Therefore, it can be inferred that recrystallization nucleation (i.e. the formation of small angle boundaries of subgrains by dislocation rearrangement) occurs during annealing. The direction distribution function is shown in figure 4f, the texture has high directivity and four main directions. Linear and curved Σ3 boundaries were found (figure 4d), but there were no annealing twins, with two parallel Σ3 boundaries. In conclusion, complete grain refinement at large undercooling has strong texture and annealing twins.

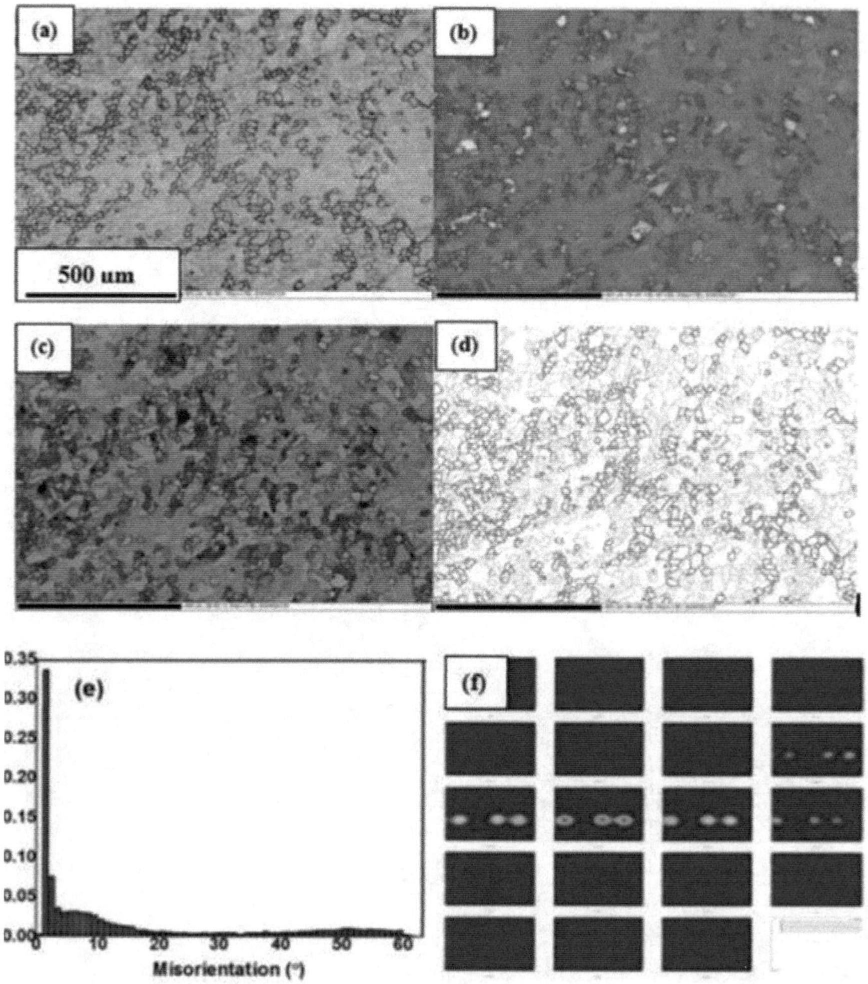

Figure 3. EBSD characterization of solidified alloy at undercooling of about 300K. (a) Comparison diagram of microstructure of solidified state and (b) EBSD diagram; (c) The microstructure is characterized by the colour represented by Euler angle; (d) Small-angle and large-angle grain boundaries; (e) Grain boundary dislocation distribution; (d) Orientation distribution function of texture.

Recrystallization is driven by deformation energy, and new deformation free grains are formed in the deformed matrix through the formation and subsequent migration of large angle grain boundaries [14]. Due to solidification of undercooled melt, the lattice distortion caused by solute capture and dendrite deformation can store deformation energy. If the grain refinement above a critical value of undercooling is due to the stress-caused recrystallization mechanism, a small angle orientation between subgrains and the subsequent large angle orientation formed by the annihilation and merger of small angle orientation would be observed.

According to Li [5], the recrystallized grain is a region with an area higher than 4 mm^2 and containing a little amount of internal misorientation, with a continuous large angle GBs along at least one third of its circumference. On this basis, a recrystallized grain can be easily distinguished by EBSD. In this study, in figures 3 and 4, many specific areas have an area > 4 mm^2 and contain a small amount of internal

IOP Conf. Series: Materials Science and Engineering 1217 (2022) 012005 doi:10.1088/1757-899X/1217/1/012005

orientation errors. However, a small portion of these specific regions have large angle GBs, while most of these specific regions have only small angle GBs. The average internal misalignment of < 1.5° in figures 3 and 4 is about 34% and 47%, respectively. Therefore, recrystallization nucleation occurs in the solidified microstructure and annealed microstructure in figure 3 and figure 4, respectively. In addition, the small angle misorientation of < 15° (small angle grain boundary) in figure 3 and figure 4 accounts for 71% and 86% of the total misorientation, respectively.

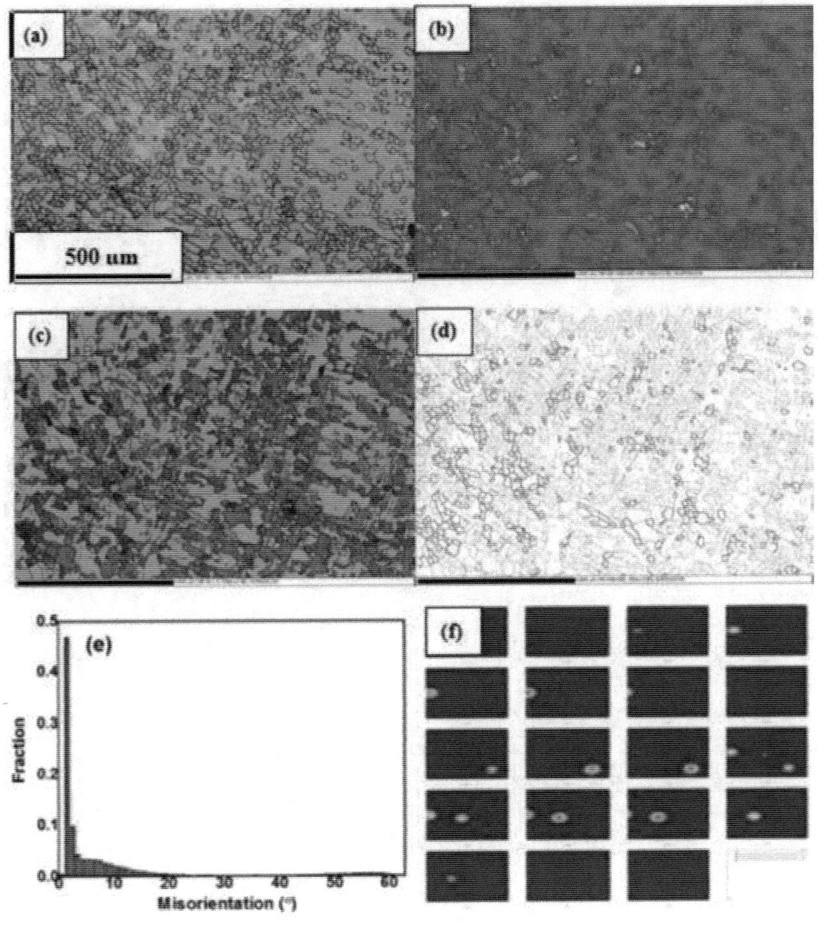

Figure 4. EBSD characterization of annealed microstructure of solidified alloy with undercooling of 300K. (a) Comparison diagram of microstructure of solidified state and (b) EBSD diagram; (c) The microstructure is characterized by the color represented by Euler angle; (d) Small-angle and large-angle grain boundaries; (e) Grain boundary dislocation distribution; (d) Orientation distribution function of texture.

Dendrite deformation can be controlled by mechanical interaction with alloy melt, that is, through fluid flow. Therefore, the predicted stress is related to the radius of dendrite tip, and the smaller the radius of dendrite tip, the greater the stress. Generally speaking, at low or medium undercooling with low solidification rate, the external fluid flow may be significant, while at large undercooling, the solidification rate is large, resulting in significant shrinkage caused fluid flow. As the driving force of dendrite fragmentation in the range of large undercooling, fluid flow plays an important role in the dendrite fragmentation process. Deformation energy is stored in dendrites and drives recrystallization.

Therefore, it can be said that dendrite fracture and recrystallization are caused by flowing fluid. Our team modelled the stress accumulated by fluid flow caused by rapid solidification shrinkage. The model shows that once dendrite coherence is formed, the strength begins to develop and exchange between dendrite skeleton, and stress gradually increases with the increase of volume undercooling.

4. Conclusions

In this paper, microstructure evolution of undercooled alloy during solidification was systematically studied and the characteristic undercooling of Ni80Cu20 alloy was confirmed. Due to non-equilibrium solidification, solidification stress is caused, which leads to dendrite deformation and strong deformation texture. Many subgrains are formed in the rapidly solidified sample, which is due to recrystallization after recalescence. The samples were further annealed at 900°C, recrystallization and dislocation network rearrangement formed new texture and subgrain. After annealing, the proportion of small angle grain boundaries increased to 86% and the proportion of large angle grain boundaries decreased to 14%. The increase of small angle grain boundary indicates that the rapid quenching alloy sample further recrystallized to form subgrains during annealing. These experimental results provide strong evidence for the recrystallization caused grain refinement in undercooled solidification of a binary alloy.

Acknowledgements

This work is supported by the Key R&D and Promotion Projects in Henan Province (No. 212102210267).

References

[1] Kramer M J and Li M 2020 Changes in short- and medium-range order in metallic liquids during undercooling. *MRS Bull.* **45** 943-50
[2] Xu X L and Liu F 2014 Observation of recrystallization upon annealing rapidly solidified microstructures of highly undercooled Ni80Cu20 alloys *J. Alloys Compd.* **597** 205-10
[3] Yang W, Xu Z F, Li W J, Cai C C, Li S, Liu F and Yang G C 2011 Comparisons of grain refinement and recalescence behavior during the rapid solidification of undercooled Cu–Co and Cu–Ni alloys. *Phys. B: Condens. Matter* **406** 3710-14
[4] Wang H, Liu F and Yang G 2011 Experimental study of grain refinement mechanism in undercooled Ni–15at.%Cu alloy. *J. Mater. Res.* **25** 1963-74
[5] Herrera Y, Grigorescu I C, Ramirez J, DiRauso C and Staia M H 1998 Micro Structural Characterization of Vanadium Carbide Laser Clad Coatings. *Surf. Coat. Technol.* **108-109** 308-311
[6] Aghaie-Khafri M and Fazlalipour F 2008 Vanadium Carbide Coatings on Die Steel Deposited by the Thermoreactive Diffusion Technique. *J. Phys. Chem. Solids* **69** 2465-70
[7] Liu N, Liu F, Yang G, Chen Y, Chen D, Yang C, Zhou Y 2007 Grain refinement of undercooled single-phase Fe70Co30 alloys. *Phys. B: Condens. Matter* **387** 151-155
[8] Lu S Y, Li J F and Zhou Y H 2007 Grain refinement in the solidification of undercooled Ni–Pd alloys. *J. Cryst. Growth* **309** 103-111
[9] Castle E G, Mullis A M and Cochrane R F 2014 Mechanism selection for spontaneous grain refinement in undercooled metallic melts. *Acta Mater.* **77** 76-84
[10] Wang H, Liu F, Yang G and Zhou Y 2010 Modeling the overall solidification kinetics for undercooled single-phase solid-solution alloys I. Model derivation. *Acta Mater.* **58** 5402-10
[11] Zhang T, Liu F, Wang H F and Yang G C 2010 Grain refinement in highly undercooled solidification of Ni85Cu15 alloy melt: Direct evidence for recrystallization mechanism. *Scr. Mater.* **63** 43-46
[12] Yang W, Liu F, Wang H F, Chen Z, Yang G C and Zhou Y H 2009 Prediction of the maximal recalescence temperature upon rapid solidification of bulk undercooled Cu70Ni30 alloy. *J. Alloys Compd.* **470** L13-L16
[13] Xu X L, Chen Y Z and Liu F 2013 Study of microstrain in rapidly solidified structures of hypercooled Co80Pd20alloys. *Mater. Sci. Technol.* **29** 117-120

IOP Conf. Series: Materials Science and Engineering 1217 (2022) 012005 doi:10.1088/1757-899X/1217/1/012005

[14] Liu J, Streufert J R, Mu K, Si T, Han T, Han Y, Lin X, Li J and Braun P V 2020 Polymer Composites Containing Phase-Change Microcapsules Displaying Deep Undercooling Exhibit Thermal History-Dependent Mechanical Properties. *Adv. Mater. Technol.* 5
[15] Xu X L, Chen Y Z and Liu F 2013 Recrystallization mechanism of grain refinement in hypercooled single phase alloys. *J. Cryst. Growth* **377** 153-159
[16] Xu X L and F Liu F 2014 Crystal Growth Due to Recrystallization upon Annealing Rapid Solidification Microstructures of Deeply Undercooled Single Phase Alloys Quenched before Recalescence. *Cryst. Growth Des.* **14** 2110-14.

Soil plasticity and standard proctor compaction characteristics of stabilized Kota Kinabalu clay using biomass silica (SH-85)

V V Vitales[1], A E Amaludin[1]*, Y Ho[1] and N A Amaludin[2]

[1] Civil Engineering Programme, Faculty of Engineering, Universiti Malaysia Sabah, 88400 Kota Kinabalu, Sabah, Malaysia
[2] Centre of Research in Energy and Advanced Materials, Faculty of Engineering, Universiti Malaysia Sabah, 88400 Kota Kinabalu, Sabah, Malaysia

*Corresponding author: adriana@ums.edu.my

Abstract. In the Pan Borneo Highway (PBH) project, the occurrence of clay soils is likely to increase construction costs, due to its low shear strength and high compressibility characteristics. In recent years, the use of non-traditional stabilizers e.g., Biomass Silica (SH-85) compound is becoming more prevalent. This study aims to determine the soil plasticity behaviour and to ascertain the Standard Proctor Compaction characteristics of clay soil by mixing it with varying dosages of SH-85. After soil classification was carried out, the Initial Lime Consumption Test was performed to estimate the suitable SH-85 dosage. The compound was mixed with the soil at concentrations of 3, 6, 9 and 15% (by soil sample weight). Moreover, Atterberg Limits test was also carried out to determine the change in soil plasticity. Subsequently, the compaction tests were performed to obtain the Maximum Dry Density (MDD) and Optimum Moisture Content (OMC) on the natural and stabilized soil. The trend of test results show that the OMC increases with increasing SH-85 dosages, while reducing the overall MDD values. The optimal quantity of SH-85 to achieve good compactibility was discovered to be 9%, producing a stabilized mix with a strength gain of +366% compared to the natural soil.

1. Introduction

The Pan Borneo Highway project is a 2,083 km long road network that connects the two states comprising of East Malaysia, namely Sabah and Sarawak. In this project, the occurrence of difficult soils e.g., clay and peat soils have been documented, and therefore this necessitates the in-depth study of this soil by the local research groups, to find novel solutions to ensure its suitability for road construction [1,2]. Weak soil occurrences in road construction are likely to increase construction costs since weak clay soil possesses poor engineering characteristics such as excessive settlements, high compressibility, and low shear strength. If not strengthened, the weak soil will eventually cause unacceptable levels of settlement and leads to increasing maintenance costs, and interruption of traffic services.

To remediate the weak soil on site, soil stabilization has been in practice in recent years, where chemical stabilization using traditional stabilizer e.g., lime and Ordinary Portland Cement is a feasible solution. Soil stabilization involves the modification of engineering and mechanical properties of soil, in order to improve weak soils to obtain predetermined values of performance [3] before the stabilized

Content from this work may be used under the terms of the Creative Commons Attribution 3.0 licence. Any further distribution of this work must maintain attribution to the author(s) and the title of the work, journal citation and DOI.
Published under licence by IOP Publishing Ltd

soil is used for construction applications. Stabilized soils possess enhanced soil properties, such as reduction of soil settlement, reducing soil compressibility, increasing soil shear strength, and improving the soil workability by modifying its compaction properties. The type of soil stabilization covered for this paper is chemical stabilization. Chemical stabilization allows the soil particle bonding to be improved via chemical means, after the stabilizer material is mixed with the weak soil and subsequently compacted [4]. Through chemical stabilization, this method is deemed more economical to solve the weak soil problem on site, compared to the replacement of weak soils with good quality fill soils. [5] states that chemical soil stabilizers are mainly classified as traditional or non-traditional stabilizers. While traditional stabilizers (e.g., lime and cement) [6, 7] have been in use for a long time, non-traditional stabilizers (e.g., polymers and silicates) [8, 9] have recently gained attention in the field of weak soil remediation. In particular, bio-derived stabilizers, which is a type of non-traditional soil stabilizer obtained from agricultural waste or biomass material, is a better choice since it produces lower greenhouse emissions compared to conventional stabilizers such as cement [10]. For this study, a bio-stabilizer called biomass silica (or known commercially as SH-85) was used as the main soil stabilizing compound. Produced by Probase Manufacturing Pvt. Ltd., the biomass silica is usually sold in 25 kilogram bags and are typically used as a soil hardening material for unpaved roads [11]. The soil stabilization effect for different types of weak soils may vary and requires further laboratory tests to establish its effectiveness.

Some existing studies pertaining to the use of SH-85 to improve soil properties have been carried out by several researchers on laterite soil [12–14], on soft clay [15,16], on marine clay [17,18] and on organic soils [19]. In the research carried out by Latifi *et al.* [12] to improve soil strength, it showed that that 9% of Biomass Silica was the optimum amount of stabilizer for laterite soil, while another study conducted by Marto *et al.* [17] that 12% of Biomass Silica was the optimum amount of stabilizer for marine clay soils. Most of these studies focused on the improvement of soil strength properties, but only two studies focused into the improvement of soil compaction properties, namely Latifi *et al.* [12] and Marto *et al.* [17]. Hence, this study was conducted to focus on the improvement of soil by examining the soil compaction characteristics of Biomass Silica stabilized soil in Kota Kinabalu, Sabah and to compare the results of this study with the findings of the stabilized soils conducted in Peninsular Malaysia. Even though the current research framework only includes the basic physical and engineering soil testing programme, the authors plan to carry out the macro-structural and micro-structural assessments of the KK Clay soil stabilized with SH-85, if the preliminary results have a promising outcome. Therefore, the limitation of this study is the exclusion of the macro-structural and micro-structural studies, which would be able to show the morphology of the stabilized soil, and this will be addressed in our future works.

In the current study, an effort was made to improve the Kota Kinabalu clay soil compaction characteristics, by determining the optimum dosage (percentage by weight) of SH-85 to obtain the desired maximum dry density. To do this, the Initial Consumption of Lime (ICL) test was carried out first, in order to set the range of SH-85 dosage used in this study. Next, the Atterberg Limit values were obtained for the stabilized soil to describe the changes in the soil plasticity of the stabilized soil, which ultimately affects the compaction behaviour of the soil mix. Lastly, the Maximum Dry Density values of the stabilized soil are correlated its soil plasticity characteristics. With the improvement in the soil plasticity and compaction characteristics, it is hoped that the weak soil in this study can be remediated and repurposed as materials used for wearing course (unpaved roads) or for subgrade layers in pavement construction and is hoped to contribute towards the use of sustainable construction materials.

2. Materials and experimental program

2.1 Clay soil

The clay soil sample used in this study was obtained from Jalan Kayu Madang, Kota Kinabalu, which is an area within the vicinity of the Pan Borneo Highway project. The soil samples were disturbed samples which were retrieved at a depth of 1.5 m from the surface and had yellowish-brown colour.

Samples used for the Atterberg Limits test were sealed to preserve the natural moisture contents prior to testing. Meanwhile for the other soil classification and compaction tests, the soil samples were air-dried under the sun, then broken into smaller pieces with a rubber mallet, before it was sieved with a 2 mm sieve. The samples passing the 2 mm sieve were collected and subsequently used for the soil classification and compaction tests. The flowchart of experimental works planned in this study is shown in figure 1.

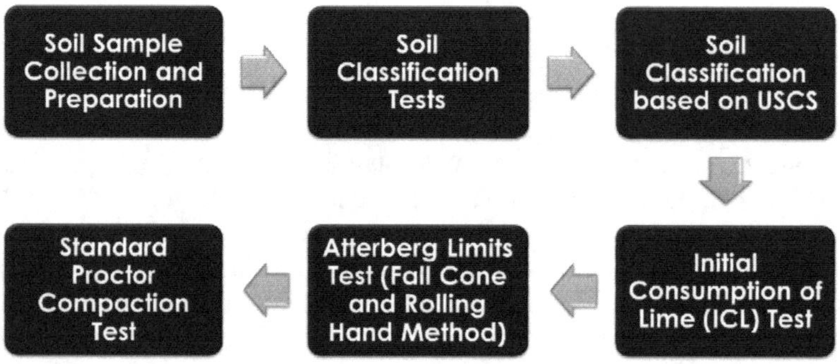

Figure 1 Laboratory tests for Kota Kinabalu clay stabilized with varying Biomass Silica content.

Table 1 shows the physical and engineering properties of Kota Kinabalu (KK) clay soil, and these obtained values were used to classify the soil according to the Unified Soil Classification System (USCS) [20]. In addition, the characteristics of KK clay soil were also compared to the findings of other Malaysian Clay soils. Based on USCS, KK Clay was classified as High Plasticity Clay, which means that the soil undergoes considerable volumetric changes with the change in moisture content and is similar to that of Rompin Clay [21]. Furthermore, according to Liu and Evett [4], this soil has medium consistency and therefore has better strength compared to the Batu Pahat and Karambunai Clay. The general aim of soil stabilization is to reduce the soil plasticity index [12, 17], to improve the soil strength [16, 17, 22] and to improve the soil compaction characteristics [3, 23]. Therefore, the Kota Kinabalu clay soil is suitable to be used for soil stabilization purposes due to its medium consistency and high plasticity.

Table 1. Engineering and Physical Properties of Untreated Kota Kinabalu Clay in comparison with other Malaysian Clay Soils.

Engineering and Physical Properties	Current Study	Laterite Clay [12]	Rompin Clay [21]	Batu Pahat Soft Clay [16]	Karambunai Clay* [24]
Liquid Limit, LL (%)	51	75	72	73	40
Plastic Limit, PL (%)	22	41	25	29	22
Plasticity Index, PI (%)	29	34	47	44	18
pH Value	3.7	5.35	5.5	-	-
Specific Gravity, Gs	2.58	2.69	2.52	-	2.61
Maximum Dry Density (kg-m^{-3})	1587	1310	1560	1343	-
Optimum Moisture Content (%)	18	34	18	30	19.6
Undrained Shear Strength, S_u (kPa)	33.08	135.00	-	16.8	15.78
Soil Classification (USCS) [20]	CH	MH	CH	MH	CL
Type of Soil	High Plasticity Clay	Inorganic Silt	High Plasticity Clay	Inorganic Silt	Medium Plasticity Clay

*NOTE: From Borehole 1 [24]

IOP Conf. Series: Materials Science and Engineering 1217 (2022) 012006 doi:10.1088/1757-899X/1217/1/012006

2.2 Biomass silica (SH-85)
Biomass silica or SH-85 is a powder-type, calcium based [12], bio-derived stabilizer but the chemical composition remains unknown since it is a commercial product. Nevertheless, Marto *et al.* [17] had performed a test with the Energy Dispersive X-Ray Spectrometer (EDAX) machine, which revealed the general chemical composition of the biomass silica. It indicated that the SH-85 material consisted mostly of calcium oxide and aluminium oxide, with a pH value of 12.65, and this finding has been confirmed by Latifi *et al.* [14]. The selected dosage of SH-85 to be used in this study ranges from 3 to 15 percent, based on past research and the range obtained from the Initial Consumption of Lime (ICL) test results that will be discussed further in section 3.1.

2.3 Sample preparation and testing programme
In addition to the engineering and physical characteristics tests, the BS 1924: Part 2: 1990 [25] standard was used to conduct the Initial Consumption of Lime (ICL) test. The ICL test is performed to obtain the minimum dosage of soil stabilizer to be used, in order to significantly improve the soil qualities. In the beaker, 20 g of weighted soil sample was added. Then, the soil sample was combined with 100ml of distilled water. After that, the pH meter was turned on and calibration solutions (pH 4.0 and pH 7.0) were used to calibrate it. Before inserting the pH metre probe into the beaker containing the soil sample solution, it was washed with distilled water. After recording the pH value, the probe was washed with distilled water. After that, 3% of SH-85, weighed by the proportion of soil (0.6 g), was then added to the beaker and agitated with a glass rod for around 3 minutes before allowing the solution to settle. The probe was then immersed in the solution, and the pH value was recorded. The procedure was repeated with the required increment of SH-85 stabiliser until the stabiliser content (by weight) reached 15%. The ICL test is shown in figure 2 (a), for soil mixes with varying SH-85 dosages.

Then, the Atterberg Limits test were conducted on the untreated and stabilized clay soil, in order to determine the change in soil plasticity as the stabilizer content increases [26]. For the Liquid Limit (LL) test, the cone penetration test is performed by mixing 500 g of soil sample with water using spatula on a glass plate. The penetration cup was filled with the mixed soil using the spatula, and the soil sample was gently pressed to remove air voids. The sample had a flat surface that was levelled with the cup's mouth. The penetration cup was then positioned beneath the cone. The cone was lowered to the point where it just brushes the surface without contacting the soil. After then, the penetration gauge was set to zero. Then, the penetration cone was released, and the penetration depth (mm) was recorded at least twice for each sample, where readings with ±0.5mm are allowed per BS 1377: Part 2: 1990: 4.3 [26]. Penetration tests are repeated until it exceeds penetration of 20 mm. To determine the moisture content value, a small sample of soil was obtained and placed in the moisture content container, which was then weighed and placed in the oven for 24 hours. The Liquid Limit is defined as the moisture value obtained at Liquid Limit value is obtained for the moisture value at 20 mm cone penetration. figure 2 (b) shows the Cone Penetration test apparatus used for this experiment.

(a) Initial Consumption of Lime test

(b) Cone Penetration Test

Figure 2 Laboratory tests for Kota Kinabalu clay stabilized with varying Biomass Silica content.

On the other hand, the rolling hand test to determine the Plastic Limit (PL) was conducted based on BS 1377: Part 2: 1990: 5.3 [26]. The rolling hand test method was conducted as follows: a soil sample was taken from the mixed soil used for the LL test. For the rolling process, a clean glass plate was prepared. The soil sample was then rolled on the glass plate, kneading the soil using by hand as the heat from the hand reduced its moisture content. The process was repeated until the soil crumbles at about 3 mm in diameter, which is about the size of a laboratory glass rod. Then, a portion of the rolled soil was placed into the moisture content can then it was weighed and placed into the oven for oven drying for 24 hours to determine its moisture content. The procedure was repeated for 2 more times to determine the average of moisture content value for the PL. The Plastic Limit value is obtained for the moisture value taken when the rolled soil sample crumbles at 3 mm diameter.

Next, a series of Standard Proctor Compaction (SPC) tests were conducted based on British Standard 1377: Part 2: 1990: 3.3 [27]. To begin, a soil sample of 2500g was weighed. The compaction mould was tightened with nuts to the collar and base plate provided. The mould was weighed, and the results were recorded. Water was added to the soil sample in the amount of 5-10% and spread on the soil before it was completely mixed with a hand roller. The soil sample was then divided into three equal portions on the tray, with the first portion placed into the mould. A 2.5 kg rammer was used to compact the soil layer for 27 blows. The next layer of soil was added in the same manner as the previous one, with the exception that the finished compacted layer should be higher than the top of the mould. The results of the SPC tests were used to obtain the optimum moisture content (OMC) and maximum dry density (MDD) of the soil sample.

3. Results and discussion

3.1 Initial consumption of lime (ICL) test

In order to determine the suitable range of Biomass Silica to be used for soil stabilization, the Initial Consumption of Lime (ICL) test was carried out based on BS 1924 [25]. As mentioned in section 2.2, the pH value of Biomass Silica is 12.65. Table 2 lists the pH value of the untreated and SH-85 stabilized KK Clay soil.

Table 2. pH value of Clay Stabilized with Various Percentage of Biomass Silica (pH value = 12.5).

Mix Proportion	High Plasticity Clay Current Study	Inorganic Laterite Silt [12]	Low Plasticity Silt [28]
Untreated Soil	3.7	4.7	5.0
Soil + 3% Stabilizer	10.7	10.6	11.3
Soil + 6% Stabilizer	11.1	-	11.5
Soil + 9% Stabilizer	11.2	11.6	11.6
Soil + 15% Stabilizer	11.4	-	11.7

In addition, the ICL test results of this study are also compared to the findings of [12] and [28]. The KK Clay soil sample has a pH value of 3.7 and is more acidic compared to the Laterite and Silt soils. However, all three types of soils exhibited the same trend after the SH-85 additive was mixed, highly acidic soils turned into an alkaline material, since the SH-85 material itself is highly alkaline. The ICL test is typically halted after there is no increase in the pH value is observed, which indicates that the soil particles have fully reacted with the stabilizer material. The authors found that the chemical reaction between the soil and stabilizer takes place immediately after mixing, and the pH value remained constant even after several hours of curing, which was also reported by [12]. At higher dosages where the stabilizer dosage exceeded the requirement for the chemical reaction to occur, this would result in the reduction of strength in the stabilized soil mix [29]. Therefore, it is important to limit the stabilizer dosage, specifically at the point where the pH value stops increasing. For this study, after the soil and stabilizer material was mixed, the pH value stopped increasing at 15% SH-85 content, with the final pH value of 11.4. As a result of the ICL test, the selected SH-85 dosages used in this study were 3%, 6%, 9% and 15%, in order to study the changes to the soil plasticity and compression characteristics of the stabilized soils.

3.2 Atterberg limits test

The Atterberg Limits test yields two values, the Liquid Limit (LL) and Plastic Limit (PL), while the Plasticity Index (PI) is obtained by subtracting the PL from the LL. In this section, the changes of Plasticity Index and the soil classification after the stabilizer is mixed with the soil is discussed. Table 3 shows the Atterberg Limits test results for the KK Clay soil mixed with varying SH-85 contents, in comparison with the studies conducted by [12] and [17] for Laterite Silt and Marine Clay mixed with SH-85. The USCS classification for the KK Clay soil sample was High Plasticity Clay (CH), with the LL, PL, and PI values of 22%, 51% and 29%, respectively. More specifically, High Plasticity soils possess LL exceeding 50 and PI exceeding 30 [30]. Figure 3 shows the Atterberg Limits of Kota Kinabalu clay stabilized at different Biomass Silica contents.

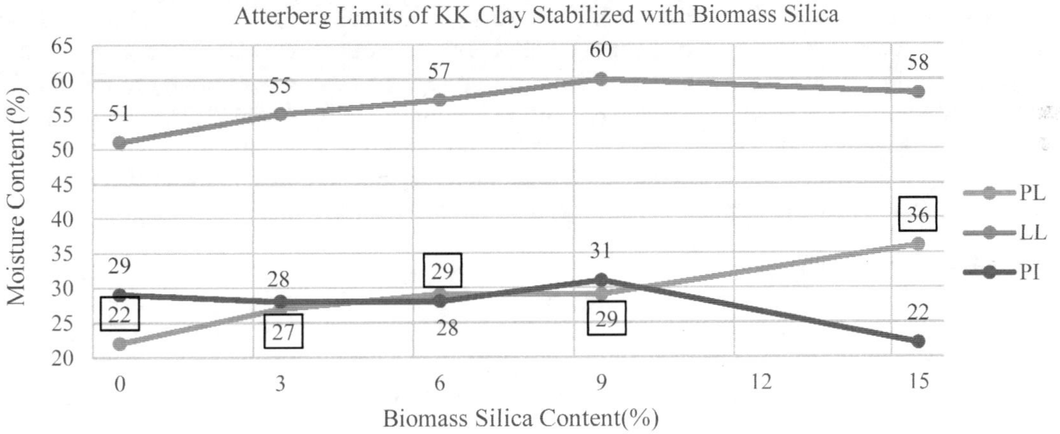

Figure 3 Atterberg Limits of Kota Kinabalu clay stabilized with varying Biomass Silica content.

Table 3. Atterberg Limits Values of Clay and Silt Stabilized with Various Percentage of Biomass Silica (compared to [12] and [17]).

Mix Proportion (Current Study)	**PL**	**LL**	**PI**	**USCS Classification**
Untreated Soil (KK Clay)	22	51	29	CH (High Plasticity Clay)
Soil + 3% Biomass Silica	27	55	28	CH (High Plasticity Clay)
Soil + 6% Biomass Silica	29	57	28	CH (High Plasticity Clay)
Soil + 9% Biomass Silica	29	60	31	CH (High Plasticity Clay)
Soil + 15% Biomass Silica	36	58	22	MH (Inorganic Silt)
Mix Proportion [12]	**PL**	**LL**	**PI**	**USCS Classification**
Untreated Soil (Laterite Silt)	40	74	34	MH (Inorganic Silt)
Soil + 3% Biomass Silica	57	89	32	MH (Inorganic Silt)
Soil + 9% Biomass Silica	62	81	19	MH (Inorganic Silt)
Mix Proportion [17]	**PL**	**LL**	**PI**	**USCS Classification**
Untreated Soil (Marine Clay)	23	59	36	CH (High Plasticity Clay)
Soil + 3% Biomass Silica	26	51	25	CH (High Plasticity Clay)
Soil + 6% Biomass Silica	27	49	21	CL (Medium Plasticity Clay)
Soil + 9% Biomass Silica	27	48	21	CL (Medium Plasticity Clay)
Soil + 12% Biomass Silica	27	48	21	CL (Medium Plasticity Clay)
Soil + 15% Biomass Silica	27	48	21	CL (Medium Plasticity Clay)

In this study, after an addition of 3% of the stabilizer, a slight increase of the LL was observed, where it increased from 51% to 55%. The LL value increased to 60% at 9% SH-85, before it decreases to 58% at 15% stabilizer content. Conversely, the PL values showed a constant increase in tandem with the increase of SH-85, where the PL value increased to 27% with 3% SH-85 content, and to 36% with 15% stabilizer content. Given these changes in PL and LL values, the Plasticity Index showed an increase up until 9% SH-85 content (PL=31), but the PI value dropped significantly at 15% SH-85, where the PI was reduced to 22%, a reduction of 7% from its natural state. Ultimately, this changed the USCS classification of the soil from High Plasticity Clay to Inorganic Silt, with improved soil plasticity characteristics due to the reduced PI value [31]. A similar trend was seen in past studies, with the Laterite Silt PI reduced from 34% to 19% due to the addition of 9% SH-85 [12], while the Marine Clay PI reduced from 36% to 21% with 6% SH-85 content [17]. In all three studies, the reduction of Plasticity Index indicates that the soil will be less susceptible to moisture-induced volume changes, while also increasing its friability [12] because the SH-85 powder modifies the soil mix into a silt-like material (crumblier like silt, and less cohesive compared to normal clay). The Biomass Silica is a calcium-based stabilizer, and the presence of calcium ions induces the coagulation and aggregation process of the clay mineral particles within the stabilized soil mix [13]. This process increases the void volume, which in turn, increases the optimum moisture content [17, 32] and this will be discussed further in Section 3.3. Similarly, as seen in the increase of the LL values as more SH-85 content was added to the KK Clay and Laterite Silt, this was caused by the increase of water holding capacity due to the coagulation process [12,17], but this LL value will only increase to a certain point called the "stabilizer fixation point" [33]. At this fixation point, the addition of more stabilizer material will not contribute to the coagulation process because the clay mineral particles are saturated by calcium cations [17].

3.3 Standard proctor compaction test

The stabilized soil specimens with 3, 6, 9 and 15% of Biomass Silica were subjected to a series of Standard Proctor Compaction tests. Figure 4 shows the relationship between the dry density and moisture contents of compacted KK Clay and SH-85 mixes, with different stabilizer dosages. Meanwhile, table 4 shows the MDD and OMC of Malaysian clay and silty soils stabilized with various types of powder-type chemical soil stabilizers. The untreated soil has an MDD of 1587 kg/m^3 and an OMC of 18%. For the addition in percentage of the stabilizer SH-85, this increases the OMC, while it reduces the MDD of the stabilized soil (at the same compactive effort). At 9% dosage of SH-85, the lowest MDD reading was obtained, where the MDD decreased to 1375 kg/m^3, while the OMC increased to 27%, this denotes a decrease of 212 kg/m^3 in the MDD and an increase of 9% in the OMC. It can be seen that the MDD values of the original KK Clay sample is much higher compared to the SH-85 treated clay, and this has also been observed by other researchers who reported a similar trend for the use of powder-type soil stabilizers with increasing stabilizer contents [17,21,32]. Marto *et al.* [13] stated that as the percentage of SH-85 increased, the decrease in MDD of the soil and an increase of the OMC was observed, due to the flocculation and agglomeration effect of soil particles (as mentioned in section 3.2). Furthermore, Zhu *et al.* [32] reported that the compaction curve of the stabilized soil becomes flatter with increasing stabilizer content, compared to the natural soil, since the stabilizer material absorbs more water required for the hydration process. This can be correlated with the LL value increase that occurs when more SH-85 content is added to the mix, which will stop at the "stabilizer fixation point", as discussed in the previous section. At the optimum stabilizer content (9% Biomass Silica content), a flatter compaction curve is seen in figure 4, and an increase in the Liquid Limit was previously seen in figure 3, where the water holding capacity increases, hence allowing more moisture to be absorbed without suffering a significant drop in the soil's dry density value and hence improving its workability on site. As a result, even if the stabilized soil has lower overall dry density, but it can be compacted more easily on site [13], with the added advantage of long-term soil strength gains as well, which will be discussed in the following paragraph.

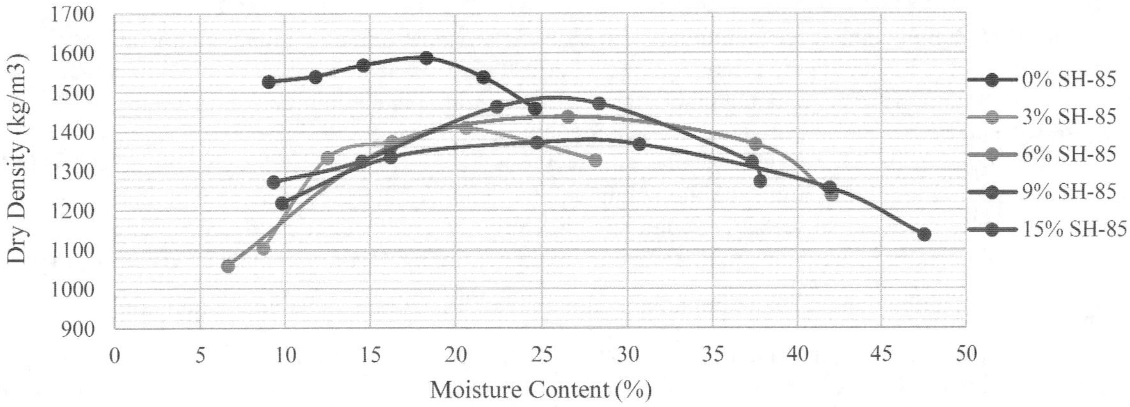

Figure 4. Maximum Dry Density and Optimum Moisture Content of KK clay stabilized with varying Biomass Silica content.

Table 4. Maximum Dry Density and Optimum Moisture Content of clay stabilized with various types of powder-type chemical stabilizers (MDD and OMC changes compared to natural soil are in brackets).

Study	USCS Soil Classification	Stabilizer Type and Soil Strength Gain	Optimum Stabilizer Dosage (%)	Maximum Dry Density (kg m^{-3})	Optimum Moisture Content (%)
Current Study East Malaysia	Kota Kinabalu Clay (High Plasticity) CH	Biomass Silica (SH-85) Lime: 68.2% Silica: 9.25% Strength Gain: +366%	9	1375 (MDD -212)	27.0 (OMC +9)
Marto *et al.* (2014) [13] West Malaysia	Johor Laterite Silt (Inorganic Silt) MH	Biomass Silica (SH-85) Lime: 68.2% Silica: 9.25% Strength Gain: +302%	9	1250 (MDD -60)	37.5 (OMC +2.5)
Zukri and Ghani (2014) [21] West Malaysia	Rompin Clay (High Plasticity) CH	Hydrated Lime Strength Gain: +5%	9	1342 (MDD -218)	23 (OMC +5)
Taib *et al.* (2016) [23] East Malaysia	Serian Silt (Inorganic Silt) MH	RBI 81 Lime: 52% Silica: 19% Strength Gain: +279%	8	1900 (MDD +160)	26.5 (OMC +3)
Zainuddin *et al.* (2019) [3] West Malaysia	Marine Clay (High Plasticity) CH	Demolished Tiles Material Calcium: 1.99% Silicon: 20.44% Strength Gain:-12%	10	1300 (MDD +100)	32 (OMC -2)

In table 4, several Malaysian clayey and silty soils are listed together with the corresponding optimal soil stabilizer dosage, and the reported soil strength gain compared to the natural soil values. Biomass Silica, used by Marto *et al.* [13] and in this study had reported an optimum stabilizer dosage of 9%, while despite recording a reduction in MDD and an increase in the OMC, had seen more than 300% increase in the soil strength. Meanwhile, Zukri and Ghani [21] found that the use of 9% Hydrated Lime yielded a minimal strength gain (5%), but the Rompin Clay was made to have better compactibility on site. Taib *et al.* [23] had managed to improve the characteristics of Serian Silt using the Road Building International (RBI) 81 stabilizer, where a 279% strength gain was recorded, alongside an increase in both the MDD and OMC values. Conversely, Zainuddin *et al.* [3] had attempted to improve the Marine Clay strength and compaction characteristics using Demolished Tiles Materials (DTM), but had instead recorded a reduction in the marine clay strength by adding 10% of the stabilizer. However, the DTM stabilizer managed to slightly improve the MDD value of the stabilized Marine Clay soil mix, and to reduce the OMC value as well. As reported in this study, and with similar trends reported by past researchers [17, 21, 23], the general observation made pertaining to the use of various powder-type stabilizer is that the lime content plays a huge role in the improvement of the soil strength and compaction properties. Lower MDD values are not a disadvantage, as the stabilized soil can be compacted easier on site while having its strength significantly enhanced at the same time. In the previous paragraph, the discussion about the Liquid Limit at the optimum SH-85 content was made, where the water holding capacity is increased, and this allows the moisture to be used to dissolve the lime contained within the SH-85 stabilizer to create more calcium ions. The presence of lime in the stabilizer materials i.e., SH-85 allows the calcium ions to react with the silica and alumina in the clay particles to form cementitious compounds, hence allowing the improvement of the soil strength and compaction characteristics [12, 13, 23, 34]. This is evident especially when we highlight the strength loss of the clay soil stabilized with DTM, since the calcium content is very low (1.99%). Therefore, it can be said that the SH-85 material, given its substantial lime content, is suitable to be used for the stabilization of KK Clay soil.

4. Conclusion

In this study, a series of soil classification and soil compaction tests were carried out to study the effect of Biomass Silica (SH-85) as bio-stabilizer material to improve the characteristics of Kota Kinabalu Clay soil. The soil sample was classified as high plasticity clay (CH) according to Unified Soil Classification System (USCS). Based on the ICL test, the suitable range of SH-85 to be used in this study was 3-15% (by soil weight), with the final pH value of 11.4 at 15% stabilizer content. Next, the Atterberg Limits test on the stabilized soil mixes showed an increasing trend in the PL and LL values, but the LL value decreased at 15% SH-85, therefore reducing the PI value, and changing the USCS Classification of the soil to Inorganic Silt (MH). Meanwhile, the overall trend of the compaction characteristics shows that the OMC tends to increase and MDD decreases with higher SH-85 content, which enables the KK Clay soil to be compacted with ease in wetter site conditions. With 9% SH-85 content, the MDD value is 1375 kg/m^3 while the OMC is 27%, but this produces a stabilized mix with a soil strength gain of +366% compared to the natural soil. The reduction in MDD and increase in OMC is caused by the flocculation and agglomeration effect of the soil particles, due to the presence of cations in the stabilized soil mix.

As previously mentioned, when the SH-85 content is increased until its stabilizer fixation point is reached, this produces a soil-stabilizer mix with a flatter compaction curve. At the optimum stabilizer content (9% Biomass Silica content), a flatter compaction curve was seen in figure 4, and the increase in the Liquid Limit was seen in figure 3, where the water holding capacity increases, hence allowing more moisture to be absorbed without suffering a significant drop in the soil's dry density value. Consequently, the increase in the water holding capacity also allows the moisture to be used to dissolve the lime contained within the Biomass Silica to create more calcium ions. The presence of moisture promotes the reaction between these calcium ions with the silica and alumina in clay particles, and this reaction is responsible for the formation of cementitious compounds, which contributes to the improved soil strength characteristics. When added in small dosages, the Biomass Silica material shows improvement to the soil plasticity and compaction characteristics of KK clayey soil and is deemed a viable stabilizer material. The recommended dosage of Biomass Silica (SH-85) in order to achieve the desired compaction characteristics on site is 9%. At the time of writing, the authors are currently planning further experimental works to study the effectiveness of Biomass Silica as a stabilizer material for other problematic soils such as organic soil.

Acknowledgement
The authors gratefully acknowledge the support from the Faculty of Engineering, Universiti Malaysia Sabah (UMS) for providing the facilities required to undertake the experimental works carried out in this project.

References

[1] Makinda J, Gungat L, Rao N S V K and Sulis S 2018 Compressibility behaviour of Borneo tropical peat stabilized with lime-sand column *Int. J. Adv. Sci. Eng. Inf. Technol.* **8** (1) 172–7
[2] Sapar N I F, Matlan S J, Mohamad H M, Alias R and Ibrahim A 2020 A Study on Physical and Morphological Characteristics of Tropical Peat in Sabah *Int. J. Adv. Res. Eng. Technol.* **11** (11) 542–53
[3] Zainuddin N, Mohd Yunus N Z, Al-Bared M A M, Marto A, Harahap I S H and Rashid A S A 2019 Measuring the engineering properties of marine clay treated with disposed granite waste *Meas J. Int. Meas. Confed.* **131** 50–60
[4] Liu C and Evett J B 2008 *Soils and Foundations* (Upper Saddle River, NJ: Prentice Hall)
[5] Eisazadeh A 2010 *Physicochemical Behaviour of Lime and Phosphoric Acid Stabilized Clayey Soils* (Skudai: Universiti Teknologi Malaysia)
[6] Veena V, Reghunath S, Cyrus S and Abraham B M 2021 Potential application of cement amended marine clay as liner material *IOP Conf. Ser. Mater. Sci. Eng.* **1114** (1) 012014
[7] Di Sante M 2020 On the Compaction Characteristics of Soil-Lime Mixtures *Geotech. Geol. Eng.* **38** (2) 2335–44
[8] Saleh S, Ahmad K, Mohd Yunus N Z and Hezmi M A 2020 Evaluating the toxicity of polyurethane during marine clay stabilisation *Environ. Sci Pollut. Res.* **27** (17) 21252–9
[9] Hillary E, Pakir F, Aziz N A A and Madun A 2021 The Effectiveness of Demolished Tile Material for Soil Improvement: A Review *IOP Conf Ser Mat. Sci. Eng.* **1144** (1) 012071
[10] Ramdas V M, Mandree P, Mgangira M, Mukaratirwa S, Lalloo R and Ramchuran S 2021 Review of current and future bio-based stabilisation products (enzymatic and polymeric) for road construction materials *Transp. Geotech.* **27** (2020) 100458
[11] Probase Manufacturing 2011 Probase Manufacturing Sdn Bhd *SH-85 Soil Hardener*
[12] Latifi N, Marto A and Eisazadeh A 2013 Structural Characteristics of Laterite Soil Treated by SH-85 and TX-85 (Non-Traditional) Stabilizers *Electron. J. Geotech. Eng.* **18 H** 1707–18
[13] Marto A, Latifi N and Eisazadeh A 2014 Effect of Non-Traditional Additives on Engineering and Microstructural Characteristics of Laterite Soil *Arab J. Sci. Eng.* **39** (10) 6949–58
[14] Latifi N, Eisazadeh A, Marto A and Meehan C L 2017 Tropical residual soil stabilization: A powder form material for increasing soil strength *Constr. Build. Mater.* **147** 827–36
[15] Kasim F, Marto A, Abdul Rahman N A and Tan C S 2015 Unconfined Compressive Strength and Microstructure of Clay Soil Stabilised with Biomass Silica *J. Teknol.* **77** (11) 1–6
[16] Nor A H M, Pakir F, Arifin A and Sanik M E 2015 Stabilization of Batu Pahat Soft Clay by Combination Between TX-85 and SH-85 Stabilizers *J. Adv. Res. Mat. Sci.* **14** (1) 1–7
[17] Marto A, Mohd Yunus N Z, Pakir F, Latifi N, Mat Nor A H and Tan C S 2014 Stabilization of Marine Clay by Biomass Silica (Non-Traditional) Stabilizers *Appl. Mech. Mat.* **695** 93–7
[18] Jin L J, Mohd Yunus N Z, Hezmi M A, Rashid A S A, Marto A, Kalatehjari R, Pakir F, Mashros N and Ganiyu A A 2018 Predicting the Effective Depth of Soil Stabilization for Marine Clay Treated by Biomass Silica *KSCE J. Civ. Eng.* **22** (11) 4316–26
[19] Hassan N, Wan Hassan W H, Rashid A S A, Latifi N, Mohd Yunus N Z, Horpibulsuk S and Moayedi H 2019 Microstructural characteristics of organic soils treated with biomass silica stabilizer *Environ. Earth Sci.* **78** (12) 1–9
[20] ASTM 2017 ASTM D2487-17 *Standard Practice for Classification of Soils for Engineering*

Purposes (Unified Soil Classification System) (PA: ASTM Int West Conshohocken)
[21] Zukri A and Ghani N A A A 2014 A Study of Soil Stabilization by Hydrated Lime at Kampung Kedaik Asal, Rompin, Pahang, Malaysia *Appl. Mech. Mater.* **695** 738–41
[22] Zukri A 2013 Pekan soft clay treated with hydrated lime as a method of soil stabilizer *Procedia. Eng.* **53** 37–41
[23] Taib S N L, Striprabu S, Ahmad F, Charmaine H J and Patricia N E 2016 Investigation on Strength Development in RBI Grade 81 Stabilized Serian Soil with Microstructural Considerations *IOP Conf. Ser. Mat. Sci. Eng.* **136** (1)
[24] Roslee M R 2019 Engineering Geological Investigation on Karambunai-Lok Bunuq Landslides, Kota Kinabalu, Sabah *Malaysian J. Geosci.* **3** (2) 01–6
[25] BSI 1990 BS 1924–2. British Standard Method of the Test for Soils for Civil Engineering Purposes: Part 2, Stabilized materials for civil engineering purposes. *Br Stand Institution, London*
[26] BSI 1990 BS1377-2. British Standard Method of the Test for Soils for Civil Engineering Purposes: Part 2, Classification Test. *Br Stand Institution, London*
[27] BSI 1990 BS1377-4. British Standard Method of the Test for Soils for Civil Engineering Purposes: Part 4, Compaction-Related Tests. *Br Stand Institution, London*
[28] Laugan A 2018 *Effect of the Non-Traditional Soil Stabilizer (SH-85) on Marine Soil in Kota Kinabalu, Sabah* (Universiti Malaysia Sabah (Unpublished))
[29] Sukmak P, Horpibulsuk S and Shen S L 2013 Strength development in clay-fly ash geopolymer *Constr. Build Mat.* **40** 566–74
[30] Casagrande A 1948 Classification and Identification of Soils *Trans. Am. Soc. Civ. Eng.* **113** (1) 901–30
[31] Bell F G 1996 Lime stabilization of clay minerals and soils *Eng. Geol.* **42** (4) 223–37
[32] Zhu F, Li Z, Dong W and Ou Y 2019 Geotechnical properties and microstructure of lime-stabilized silt clay *Bull. Eng. Geol. Environ.* **78** (4) 2345–54
[33] Rahmat M N and Ismail N 2011 Sustainable stabilisation of the Lower Oxford Clay by non-traditional binder *Appl. Clay Sci.* **52** (3) 199–208
[34] Saeed K A 2014 *Physicochemical Characterization of Lime and Cement Stabilized Clayey Soils Contaminated by Heavy Metal* (Skudai: Universiti Teknologi Malaysia)

IOP Conf. Series: Materials Science and Engineering 1217 (2022) 012007 doi:10.1088/1757-899X/1217/1/012007

Shredder machine for plastic recycling: A review paper

J H Wong[1*], M J H Gan[1,2], B L Chua[1], M Gakim[1], N J Siambun[1,2*]

[1] Mechanical Engineering, Faculty of Engineering, Universiti Malaysia Sabah, Jalan UMS, 88400 Kota Kinabalu, Sabah, Malaysia.
[2] Centre of Research in Energy and Advanced Materials, Faculty of Engineering, Universiti Malaysia Sabah, Jalan UMS 88400 Kota Kinabalu, Sabah, Malaysia.

[*]Corresponding author: nancyjs@ums.edu.my, wongjinhoong@gmail.com

Abstract. Plastic pollution has become a global concern due to poor plastic waste disposal management. As an alternative, independent recycling efforts are necessary. A Plastic shredder machine is a preliminary machine used to cut plastic waste into small pieces before turning it into useful products. The concept design of the shredder machine that is currently available is fairly similar. The shaft and blades are the critical components in the shredder machine that determines its performance. The geometry and orientation of the blades that were fitted into the single or double-shafts were found to directly affect the shredding performance. Therefore, this article aims to review the various geometry and orientations of the blades that give direct effect on the shredding performance, as well as identifying the research gaps related to the shredder machine for plastic waste materials.

1. Introduction

Plastics have become an integral part of society since the 1950s when their mass production commenced and their usage has continued to increase with population growth, the standard of living and technological development [1]. The benefits of plastic are undeniable. It is cheap, lightweight and easy to mould. Due to their versatility, plastics have replaced traditional materials such as metal, wood and leather [2]. Today, plastic is widely used in food and beverage packaging, textile, automotive, manufacturing, medical devices such as surgical equipment, drips and blister packs for pills [3]. The increasing usage of plastics has caused some substantial environmental pollution burden on both land and water habitats [4].

According to the report produced by the United Nations Environment Programme (UNEP) in 2018, only 9% of the nine billion tonnes of plastic the world has ever produced has been recycled, while most end up in landfills, dumps or in the environment [5]. The plastic recycling process is commonly carried out through centralized networks [6]. Unfortunately, this can be seen as the reason for the low recycling rate because of the challenges of collection, separation, sorting processes and transportation of the plastic waste. In developing countries, including Malaysia, there is no adequate waste management infrastructure for recycling [7]. In addition, the process requires significant capital and extensive operating investment [8].

The importance of plastic recycling however has not gone unnoticed, independent companies, NGO's and local communities are forced to make their own efforts to recycle plastic where possible. One example is the non-profit Australian NGO called Sea Communities, which has initiated a recycling centre for the community in Les Village located on the northern coast of Bali Indonesia to help solve

Content from this work may be used under the terms of the Creative Commons Attribution 3.0 licence. Any further distribution of this work must maintain attribution to the author(s) and the title of the work, journal citation and DOI.
Published under licence by IOP Publishing Ltd

problems with waste plastic [9]. In the plastic recycling process, the primary machine used is a shredder machine. The machine is used to cut waste plastic into tiny pieces for easy handling, easier to melt and turn into other products. In small to medium scale recycling, a custom-made shredder machine was designed and fabricated [10]. A variety of designs of small to medium-scale custom-made shredder machines has been documented. This article aims to review the (i) type of recyclable plastic waste that is suitable for the small to medium-scale shredder machine (ii) the design of various shredder machines, and the (iii) performances of the shredder machine.

2. Recyclable plastic waste

Plastic products usually consist of high molecular weight polymer that may be incorporated with additional substances such as plasticizer, fillers, antioxidants, flame retardants, colouring etc. to impart desired functionalities, thus creating a vast variety of plastic materials of differing properties [11, 12]. There are two types of plastic: thermoplastic and thermoset. However only the thermoplastic is recyclable. Thermoplastic molecules are held together by weak bonds. When enough heat is applied, the plastic material can be repeatedly softened and reshape to new products [12]. Recyclable plastic usually comes with a recycling numbered symbol to assist consumers to segregate waste plastic to specific types. This code was developed by the American Society of Plastics Industry [13]. Seven (7) plastic types are; (1) Polyethylene Terephthalate; *PET* or *PETE*, (2) High-Density Polyethylene; *HDPE*, (3) Polyvinyl Chloride; *PVC*, (4) Low-Density Polyethylene; *LDPE*, (5) Polypropylene; *PP*, (6) Polystyrene; *PS* and (7) other types of plastics which do not fit in any of these categories [13].

The National Solid Waste Management Department (JPSPN) of Malaysia has surveyed to estimate the post-consumer plastic waste generation in Malaysia for 2016. Table 1 shows the estimated amount and type of plastic waste generated in major areas/states in Malaysia [14]. The top 3-types of plastic that have contributed to the waste plastic were LDPE, HDPE and PET.

Table 1. Post-consumer plastic waste generation in Malaysia, 2016 estimations [14].

	Plastics (tonnes)							
Region	**LDPE (Type 4)**	**HDPE (Type 2)**	**PET (Type 1)**	**PS (Type 6)**	**PP (Type 5)**	**PVC (Type 3)**	**Other Plastic (Type 7)**	**Total**
Northern	58,317	52,467	36,241	20,600	20,884	6,399	1,919	**196,788**
Southern	49,774	45,383	26,683	14,906	18,508	3,722	922	**159,898**
Klang Valley	118,108	113,538	67,402	37,720	42,123	11,174	1,300	**391,365**
East Coast	41,351	32,674	24,334	16,526	13,458	5,274	864	**134,482**
Sarawak	25,460	23,024	14,164	9,723	8,758	2,291	407	**83,828**
Sabah & Labuan	30,359	26,398	19,543	13,770	8,954	3,861	819	**103,703**
Total	**323,370 (30.1 %)**	**293,485 (27.4 %)**	**188,366 (17.6 %)**	**113,245 (10.6 %)**	**112,645 (10.5 %)**	**32,721 (3.1 %)**	**6,231 (0.7 %)**	**1,070,064**

However, in Malaysia, the rigid mono materials of HDPE, PET and PP plastic waste are mainly collected for recycling. Unlike flimsy plastic, rigid mono-material such as drinking bottles are easier to collect, have a higher weight per piece, less contaminated and only requires sorting by colour, crushing and washing before processing. These 3 types of plastic waste have a strong existing market in Malaysia with a complete local supply chain for collection, aggregation and processing [14].

It was also noticeable that most independent companies, NGOs or local communities have also chosen similar types of plastic waste to be recycled, probably due to plastic waste availability and the less pre-processing requirements. Thus, the designs of plastic shredder machines are usually targeted for these plastic types. Important properties of different plastic types are listed in table 2. Plastic with higher mechanical properties will induce higher stress to the shredder machine.

Table 2. Mechanical Properties of Various Types of Plastics [15].

Material	Density (g/cm^3)	Modulus of Elasticity (GPa)	Yield Strength (MPa)	Tensile Strength (MPa)
Low Density Polyethylene (LDPE)	0.925	0.172 – 0.282	9.0 – 14.5	8.3 – 31.4
High Density Polyethylene (HDPE)	0.959	1.08	26.2 – 33.1	22.1 – 31.0
Polyethylene Terephthalate (PET)	1.35	2.76 – 4.14	59.3	48.3 – 72.4
Polystyrene (PS)	1.05	2.28 – 3.28	-	35.9 – 51.7
Polypropylene (PP)	0.905	1.14 – 1.55	31.0 – 37.2	31.0 – 41.4
Polyvinyl Chloride (PVC)	1.30 – 1.58	2.41 – 4.14	40.7 – 44.8	40.7 – 51.7

3. Shredder machine

In general, the design concept of the shredder machine is somewhat similar. Every design consists of: (i) a shaft and sets of blades, (ii) a motor, gearing, bearing and transmission (iii) hopper and (iv) a framework. However, design (i) which is the combination of shaft and the sets of the blade, is the most complex component of the shredder machine because it affects the machine efficiency. Moreover, it also requires regular maintenance due to wear and tear during the shredding process. Notably, the design of the shredder machine may have a single or double-shaft. Apart from that, a variety of blade geometry and configurations/orientations of the blades in the shaft have been reported. Thus, the design related to the shaft, blades geometry and orientation of blades were reviewed in the following sections.

3.1 Single and double-shafts shredder machine

The shaft in the shredder machine is fitted with an arrangement of blades and knives to perform the cutting. A heavy-duty or large shredder machine may be necessary to have more than one shaft to increase the cutting actions and speed up the process. However, it is adequate to have a single-shaft or double-shaft for a small to medium-scale shredder machine. Several designs of small to medium-scale shredder machines have been documented. For example, Tegegne *et al.* (2019) designed a double-shafted plastic shredder. The shafts were fitted with eighteen blades with a spacer in-between [16]. Each blade has 3-teeth with a cutting angle of 56°. The arrangement/orientation of the blades on the shaft was not mentioned or shown. The two shafts rotate in opposite directions to each other. A triangle protruding fixture is attached to the side of the shredder box to close the gap between blades and to push plastic material into the gap between the rotating blades. An example of a double-shafts shredder machine is shown in figure 1 (a).

Ekman (2018) adopted a shredder machine design from the Precious Plastic group [18]. The single-shaft plastic shredder was invented by Dave Hakkens and the design is available on the Precious Plastic website [10]. The single-shaft is fitted with fifteen *s*-shape blades (5mm thickness) and a spacer placed in-between the blades. The arrangement/orientation of the blades on the shaft is spiral. Fixed knives are placed on the side of the shredder box. The cutting action occurs when the single-shaft rotates and the plastic material is cut by the rotating blades and fixed blades. An example of a double-shafts shredder machine is shown in figure 1 (b).

In prior mentioned designs, both shredder machines were fitted with sets of blades, arranged in specific orientations. An interesting design was reported by Biddinika, *et al.* (2017), where the single-shaft consisted of three-longitude blades as shown in figure 1(c), with two static blades attached to the frame. The shredding process occurs when the position of the rotating blades coincides with the static blades. However, it has been reported that the shredding blades were quickly worn out [19].

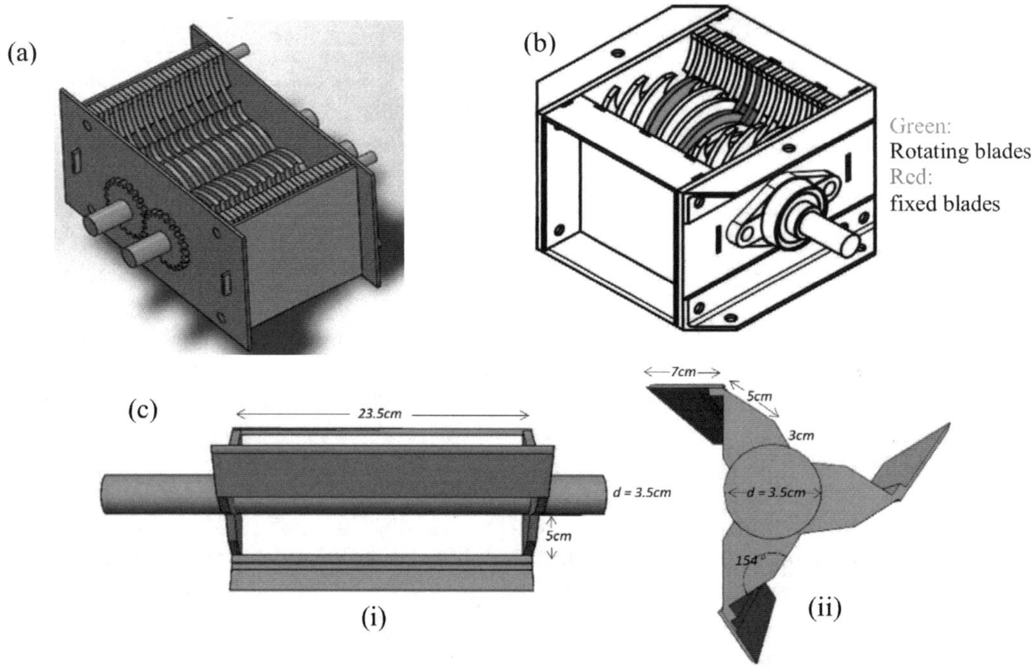

Figure 1. (a) Double shafts [17], (b) Single shaft [18], and (c) Single-shaft with horizontal blade (i) side view and (ii) front view [19].

3.2 Blades geometry.
The blades used in the shredder machine have variations of geometry. In this article, the geometry of the blades refers to the number of cutting edges (teeth/hooks), cutting angle, grabbing surface and thickness of cutter blades. If the blades are not suitable, the shredder machine may not be able to shred tougher objects, or the machine may experience clogging thus forcing a reverse of the spin direction.

3.2.1 Number of cutting edge. The number of cutting edges or teeth varies the point of contact at the time of cutting and the distribution of the forces along with the piece, thus directly affecting the performance of the machine [20]. According to Ekman (2018), the increased number of teeth increased the cutting, however may also increase the risk of material skipping on top of the blade [18]. This was also confirmed in the fabrication and test run by Rathnam and Babu (2017). Sets of six-edged blades (figure 2a) have been used in dual-shaft shredder machines. The finding concluded that the greater number of edges do not provide a better shredding performance but have raised the chance for objects skipping and winding up on top of the cutting blade [21].

Ravi (2018) found that usage of multi-edged blades has a drawback, where the blades neither cuts nor ignores the material without cutting [22]. Moreover, when shredding hard material, the blades tend to reverse against the shaft rotation before rotating and striking again on the same surface. Thus, the researcher had changed from using multi-edge blades to two-edged blades (figure 2b). The two-edged blades are found to work efficiently, with decreased working time. Other inventors and researchers have used different numbers of blade edges. For example, VijayAnanth *et al.* (2018) used a triple-edged blade (figure 2c) in a double-shafted shredding machine. The triple-edged blades reportedly also work well and can shred plastics [23].

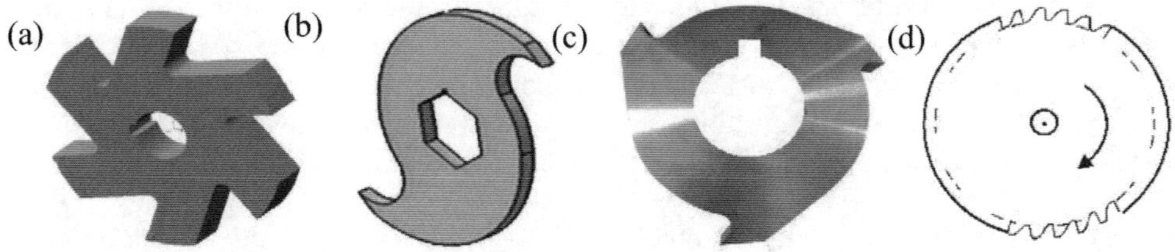

Figure 2. (a) Six-edges blade [21], (b) Two-edges blade [22], (c) Three-edges blade [23], and (d) Saw-teeth-blades [24].

An interesting blade design was used by Raji *et al.* (2020). Instead of having blades with a specific number of edges, each blade has PV-form teeth (figure 2d). The blades were fitted to a double shaft that rotates in either direction. Trial runs to shred plastic materials have shown satisfactory cutting performance for HDPE, PVC and PET types of material, but lower cutting efficiency for LDPE types of materials due to the poor flexural strength and flexural modulus of the LDPE. The increase in shaft rotation speed has further reduced the efficiency [24].

3.2.2 Cutting angle and grabbing surface of blades. Another important feature of the blade is the cutting angle, which is defined as the angle between the hook/teeth and the blade main structure as shown in figure 3. A larger cutting angle provides a larger grabbing area, thus giving a better cutting performance at greater shredding force. The reported work did not emphasize the cutting angle. Nevertheless, the cutting angle ranging from 50° to 80° were widely used for the shredder blades [16, 25-27]. There are no extensive studies carried out to determine the effects of the cutting angle to the performance of plastic shredding. Very often, the angle was chosen based on the blade size and the number of cutting edges. The grabbing surface is important in providing sufficient grabbing area on the material to be shredded to avoid uncut material. There are two types of grabbing surface: a curvature and flat grabbing surface as shown in figure 4 (a) and (b) respectively.

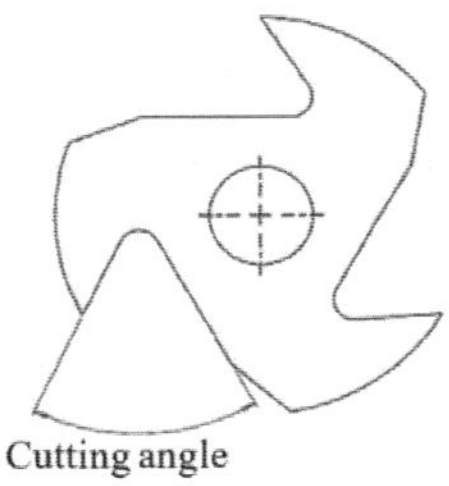

Figure 3. Schematic Diagram of triple-edge blade showing the cutting angle.

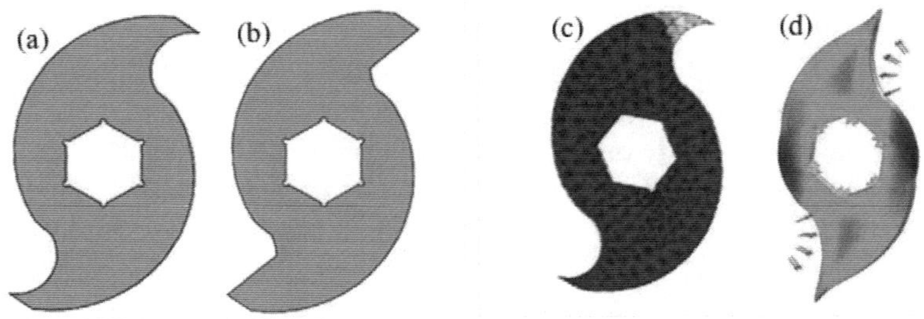

Figure 4. Shredder blades with (a) curvature and (b) flat grabbing surface, and the stresses experienced by the double-edges blade with (c) curvature [22] and (d) flat [20] grabbing surface.

So far there are no reported findings that associate the grabbing surface to the shredding performance of plastic materials. Nevertheless, it is evident that high stress and strain were experienced on the tip edge of the blade with curvature grabbing surface [22] as shown in figure 4 (c), while the stress tends to be maximum on the flat surface of the grabbing area [20] as shown in figure 4 (d). Consequently, these areas are expected to experience wear after a period and require replacing.

3.2.3 Thickness of blades. In designing a suitable blade to be used in a shredder machine, the thickness of the blades should be taken into consideration. The thickness of blades affects the shredding force, the stress concentration on the cutting edges and the safety factor of the blades. Besides that, it also affects the specific cutting energy and shredding efficiency of the shredder [28]. Dave Dakken, the pioneer of precious plastic committee claimed that thicker blades are stronger and produce bigger flake output [10]. Finite Element Analysis with ANSYS was implemented for calculating the induced stresses and strains throughout the structure of the blades [22, 25-27]. The detailed design feature of blades and the structural analysis including deformation, shear stress (equivalent to Von Mises stress) and shear strain for various thicknesses and design of blades in several researchers are tabulated in the following table 3.

A comparison on the maximum shear stress and strain, that were experienced by all four blades could not be done fairly, because of the varies variety of factors or parameters being used, for example, the building materials, diameter, cutting edge, shaft number, and the number of blades used varies. However, the data obtained theoretically can be used as a reference or comparison for future research. The thickness and size of the blade would depend on the size of the shredder machine and the type of material to be shredded. Systematic research needs to be carried out to determine the effect of blade thickness on the shredder machines' performance and the shelf-life of the blade.

Table 3. Results Obtained by Structural Analysis in Various Designs of Shredder Blades.

Properties	**Ravi (2018) [22]**	**Akash *et al.* (2019) [25]**	**Nasr and Yehia (2019) [27]**	**Kumaran *et al.* (2020) [26]**
Building material of blade	Tool Steel with Chromium or Nickel coating	Mild steel	Mild steel	Cast iron blade with the nickel coating.
Blade Thickness (mm)	5	7	10	20
Diameter (mm)	100	184	120	150
Cutting-Edge	2	4	3	3
Shaft Number	Single	Double	Double	Single
No. of blade/shaft	14	8	-	9
Maximum Von Mises Stress (Pa)	2.4734×10^6	360.05×10^6	201×10^6	164×10^6
Elastic Shear strain	1.24×10^{-5}	4.68×10^{-3}	1.40×10^{-3}	1.02×10^{-5}
Deformation (mm)	4.0321×10^{-6}	3.0727×10^{-4}	0.97×10^{-5}	3.78×10^{-5}

3.3 Orientation of blades

In a shredder machine, blades are mounted on the rotatory single-shaft or double-shafts, with spaces placed in between the blades. During operation, the shredder generally directs the material to be shredded between two sets of blades arranged along with two parallel axles. Each set of blades are interlaced with overlapping radii of the opposite axes. Cutting is performed at the interface of the two opposing sets of blades [29]. The blades can be arranged to specific orientation [18, 22, 26, 30]. Examples are shown in figure 5.

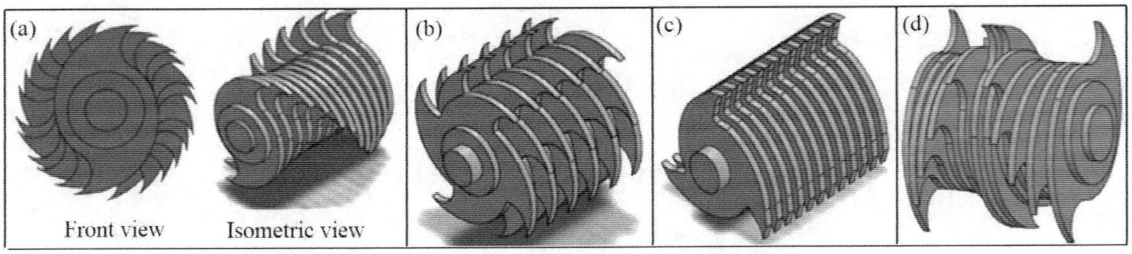

Figure 5. Rotating shaft with **(a)** Two-edges spiral, **(b)** Four-edges zig-zag, **(c)** Three-edges series, and **(d)** Two-edge V-shape orientation. Redrawn images of [22], [30], [26], [18].

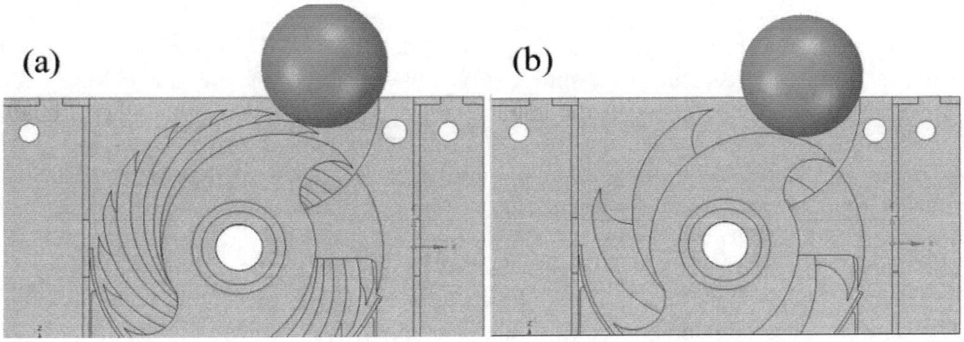

Figure 6. Shredder machine with blade arrangement in **(a)** Spiral, and **(b)** Non-spiral [31].

In the forum discussed in the Precious Plastic Committee, problems were encountered in a single-shaft shredder machine [10, 31]. Blades with spiral orientation tend to experience skipping of material at the top of the blades. Apart from that, it was also reported that the plastic waste experienced melting due to the heat built up from compression when the material shifted to one side due to the spiral blade orientation. However, changing the blade orientation was claimed to solve the problem [31]. Figure 6 shows two-blade orientations in a single-shaft shredder machine. In situation (a) the blades touch the plastic and prevented the next blade from catching it, however in (b), the next blades were able to catch the material.

The shredded particle size is dependent on the cutter-to-cutter clearance or the gap between overlapping blades. Thus, for double-shafted shredder machines, the cutter clearance is the gap between the cutter of the two rotating shafts, while in a single-shaft shredder machine, the cutter clearance is the gap between the rotating shaft and the fixed cutter. The gap between overlapping blades is set to optimal to obtain the most effective shearing. If the gap is too big, the plastic material tends to get stuck between blades and spacers, thus building friction. Thus, to obtain the specific size of the plastic pellet after shred, the spacing between the blades and the spacers are the main parameters to be adjusted [16, 32].

4. Performance of shredder machine

One of the key indicators to the success of designing and fabricating a shredder machine is by measuring its performance. Researchers have reported ways of determining the performance of their shredder machine design. One researcher, determined the shredding performance by measuring the (i) Specific Mechanical Energy (*SME*), (ii) The throughput (*TP*), and (iii) Recovering Efficiency (*RE*) [33].

The *SME* is a ratio of Energy which is the Power (W) x Time (t) to the Output mass (Q) as indicated by equation 1. The lower the *SME*, the better the shredder machine work because less energy is used for higher shredding output. The *TP* is a ratio of the output of recycled waste plastic (Q) to the time taken for recycling (t) as indicated by equation 2. Ideally, a higher TP is obtained by the shredding process. Lastly, the *RE* percentage is the ratio of Output mass (Q) to the Input mass (I) of plastic. The formula is shown as equation 3. A higher percentage indicates a good shredding process [33, 34].

$$Specific\ Mechanical\ Energy = \frac{Power\ (P) \times Time\ (t)}{Output\ mass\ (Q)} \quad (1)$$

$$Throughput\ (TP) = \frac{Output\ mass\ of\ recycled\ waste\ plastic\ (Q)}{Time\ taken\ for\ recycling\ (t)} \quad (2)$$

$$Recovering\ Efficiency\ (RE) = \frac{Output\ mass\ of\ recycled\ waste\ plastic\ (Q)}{Input\ mass\ of\ waste\ plastic\ (I)} \times 100 \quad (3)$$

$$M_R = I - Q \quad (4)$$

$$Retention\ (\%) = \frac{M_R}{M_I} \times 100\% \quad (5)$$

The performance of the shredder machine may also be determined by the retained mass (M_R), or also known as plastic that was left uncut in the shredder machine [34]. The formula is given as equation 4, where the M_R is the difference between the Input mass of waste plastic (I) and the Output mass of recycled waste plastic (Q). Of course, the lower the M_R value, the better the shredding process. The percentage of the retention is determined using equation 5, where M_I is the input mass [33, 34].

5. Conclusion

In Malaysia, the highest number of plastic wastes is of LDPE type, followed by HDPE, PET, PS, PP and PVC. The rigid mono-material of HDPE, PET and PP waste is preferable to be recycled due to easier collection and less contamination. Shredder machines can be used for turning waste plastic materials into flakes or smaller pieces. The performance of the shredder machine relies on the shaft and blades. Sets of blades are fitted to the shaft, thus producing a rotating blade during the shredding process. In small to medium shredder machines, single and double shafts may be sufficient. However, the blade geometry and the orientation of the blade on the shaft could be directly affected by the ability of the blade to grab and cut the material that needs to be shredded. Up to now, there are no systematic studies have been carried out to compare and analyse the effects of the blade's geometry and the orientation of the blade on the shaft. Studies on these areas need to be carried out to increase the shelf-life of the shredder machine's blades.

Acknowledgment

This work was supported by the Ministry of Higher Education, Malaysia (FRGS/1/2020/TK0/UMS/02/17)

References

[1] Barnes D K, Galgani F, Thompson R C and Barlaz M 2009 Accumulation and fragmentation of plastic debris in global environments *Philosophical Trans. of the Royal Ssoc. B: biological Sci.* **364**(1526) 1985-98

[2] Sivan A 2011 New perspectives in plastic biodegradation *Current opinion in biotechnology* **22**(3) 422-6

[3] Chen H L, Nath T K, Chong S, Foo V, Gibbins C and Lechner A M 2021 The plastic waste problem in Malaysia: management, recycling and disposal of local and global plastic waste *SN Appl. Sci.* **3**(4) 1-15

[4] Thompson R C, Moore C J, Vom Saal F S and Swan S H 2009 Plastics, the environment and human health: current consensus and future trends *Philosophical Trans. of the Royal Soc. B: Biological Sci.* **364**(1526) 2153-66

[5] United Nations Environment Programme UNEP *Report on Single-Use Plastics* 2018

[6] Santander P, Sanchez F A C, Boudaoud H and Camargo M 2020 Closed loop supply chain network for local and distributed plastic recycling for 3D printing: a MILP-based optimization approach *Resources, Conservation and Recycling* **154** 104531

[7] D'Ambrières W 2019 Plastics recycling worldwide: current overview and desirable changes *Field Actions Science Reports The journal of field actions (Special Issue 19)* 12-21
[8] Kreiger M A, Mulder M, Glover A G and Pearce J M 2014 Life cycle analysis of distributed recycling of post-consumer high density polyethylene for 3-D printing filament *J. of Cleaner Production* **70** 90-6
[9] Sea Communities 2020 *Plastic for Profit* 2020 [Available from: https://seacommunities.com/plastic-profit]
[10] Precious Plastic Universe 2020 *Precious Plastic* 2020 [Available from: https://preciousplastic.com/]
[11] Ikechukwu O 2017 Design of used PET bottles crushing machine for small scale industrial applications *Int. J. of Eng. Tech.* **3**(3) 157-68
[12] Halden R 2010 Plastics and public health *Annu Rev Pub Health* **31** 179-94
[13] Özkan K, Ergin S, Işık Ş and Işıklı İ 2015 A new classification scheme of plastic wastes based upon recycling labels *Waste Management* **35** 29-35
[14] WWF-Malaysia *Study on EPR scheme assessment for packaging waste in Malaysia* 2020
[15] Callister W D and Rethwisch D G 2018 *Materials science and engineering: an introduction* (New York: Wiley)
[16] Tegegne A, Tsegaye A, Ambaye E and Mebrhatu R 2019 Development of dual shaft multi blade waste plastic shredder for recycling purpose *Int. J. of Res. and Sci. Innov. (IJRSI)* 6
[17] Sakthivel M, Rajeshkannan G, Naveenkumar M, Muralimanokar M 2017 Design and analysis of twin shaft shreader using Pro-E and hyperworks software *Int. J. of Adv. Res. in Basic Eng. Sci. and Tech. IJARBEST* **3**(24) 836-58
[18] Ekman R 2018 *Development of a plastic shredder Msc Theses* (Sweeden: LUND University)
[19] Biddinika M K, Syamsiro M, Hadiyanto A N, Mufrodi Z and Takahashi F 2017 Technology for public outreach of fuel oil production from municipal plastic wastes *Energy Procedia* **142** 2797-801
[20] Yepes C P, Pelegrina R M A and Pertuz M G J 2018 Analysis by means of the finite element method of the blades of a PET shredder machine with variation of material and geometry *Contemporary Engineering Sciences* **11**(83)
[21] Rathnam A V and Babu U H 2017 Optimal design and analysis of twin shaft shredder *Innovation (IJRI)* **4** 805-13
[22] Ravi S 2018 *Utilization of Upgraded Shredder Blade and Recycling the Waste Plastic and Rubber Tyre* 2018 Proceedings of the International Conference on Industrial Engineering and Operations Management Paris, France
[23] VijayAnanth S, Sureshkumar T, Dhanasekaran C and Kumar A 2018 Design and fabrication of plastic shredder machine for clean environment *Int. J. Management, Tech. And Eng.* 8
[24] Raji A, Kuku O, Ojo S and Hunvu M 2020 Design development and performance evalutation of waste plastic shredder *J. of Production Eng.* **23**(1) 22-8
[25] Akash B P, Christina, S D K and Manoj 2019 Plastic waste management by mechanical shredder machine *IJARIIEE* 5(2) 3218-23
[26] Kumaran P, Lakshminarayanan N, Martin A V, George R and JoJo J 2020 *Design and analysis of shredder machine for e-waste recycling using CATIA* (IOP Conference Series: Materials Science and Engineering: IOP Publishing)
[27] Nasr M F and Yehia K A 2019 Stress Analysis of a Shredder blade for cutting waste plastics *J. of Int. Soc. for Sci. and Eng.* **1**(1) 9-12
[28] Sridhar N and Surendrakumar A 2017 Performance evaluation and modification of shredder cutting mechanism *Int. J. of Agric. Sci. and Res.* **7**(5)
[29] Dattatraya R P and S M Bhalgat 2020 Cutter blade design for shredder machine *Int J Sci Res Publ* **10**(06) 1022-7

[30] Siddiqui F, Patil H, Raut S, Wadake O and Tandel S 2017 Design and fabrication of paper shredder machine *Int. Res. J. Eng. and Tech. (IRJET)* **4**(6) 770-5

[31] Dave Hakkens 2020 *Forum: Shredder design has major flaw* 2020 [Available from: https://davehakkens.nl/community/forums/topic/shredder-design-has-major-flaw/]

[32] Hopewell J, Dvorak R and Kosior E 2009 Plastics recycling: challenges and opportunities *Philosophical Trans. of the Royal Soc. B: Biological Sci.* **364**(1526) 2115-26

[33] Ayo A W, Olukunle O and Adelabu D 2017 Development of a waste plastic shredding machine *Int J Waste Resour* **7**(21-4)

[34] Oyebade D, Okuluna O and Olanrewaju O 2019 Development of shredding and washing machine for polyethylene terephthalate (PET) bottles pelletizer *Int. J. Eng. Sci. and Appl.* **3**(2) 102-9

Fabrication and preliminary assessment of aramid reinforced polymer composite material

N H M Siti and W H Choong*

Mechanical Engineering Programme, Faculty of Engineering, Universiti Malaysia Sabah, Jalan UMS, 88400 Kota Kinabalu, Sabah, Malaysia.

*Corresponding author: whchoong@ums.edu.my

Abstract. A knowledge of the material constituents and fabrication process is generally required for easing the composite material properties study and characterization. This paper presents the methodology of aramid reinforced polymer composite material preliminary assessment and fabrication. The demonstration is carried out through the development of mechanical properties characterization specimen. The composite material specimen is fabricated by utilizing the open mould with hand lay-up method, where three types of laminates include of unidirectional, orthotropic, quasi-isotropic cross plies are demonstrated. A template is applied for preparing specific fibre orientation 45° angle. Plies are lay accordingly with guided by a square guide for minimizing the off-orientation angle defect. Preliminary assessment is included of verification of unidirectional aramid fabric quality, cured specimen physical defect, internal defect through micrography analysis and material volume fraction prediction. Average specimen fibre volume fraction of 0.64 is predicted by determined the fibre filament geometry detail through micrography analysis. There are several fabrication defects had been identified. The defect identification findings shall be referred for further improvement of material preparation and fabrication method.

1. Introduction

Aramid fibre had been widely used as the reinforcement material for composite material development for aircraft, aerospace, and military-related application. The aramid fibre posed a low density, high tenacity, and high modulus properties, where it promoted as one of the most excellent reinforcement materials in composite material development [1]. The excellent of its strength related property is due to its molecular structure construction and aligned to the fibre axis. The molecular structure is a long-chain synthetic polyamide with two aromatic rings attachment [2]. Yet, the aramid fibre main weakness is the interface of the material where it exhibits poor adherence and divergence due to being extremely smooth and chemically inert surface. Therefore, studies in determining the compatibility of fibre-matrix preference, fibre orientation, various treatment plan, matrix modification procedure, related composite material fabrication, and so forth have been focused [3, 4].

There are numerous composite material fabrication methods, either conventional or automated production. Automated or semi-automated fabrication such as injection moulding had the flexibility and tolerability for the changes in product design and production conditions [5]. The present work focuses on the conventional polymer-based composite materials fabrication, which adopting the hand lay-up lamination method. The hand lay-up lamination method is a conventional hand laminate the dry fabrics with resin. It is also known as wet lay-up, where each of the plies is coated with resin and laminated on

Content from this work may be used under the terms of the Creative Commons Attribution 3.0 licence. Any further distribution of this work must maintain attribution to the author(s) and the title of the work, journal citation and DOI.
Published under licence by IOP Publishing Ltd

the mould surface. The lamination will be debulked with a roller for consolidating the laminate and removes trapped air bubbles for avoiding formation of voids, which can weaken the composite material [6]. Debulking methods such as vacuum debulking with heat or no heat is an effective and quality assurance option compared to manual debulking method.

Interfacial mechanics are important in discussing material behaviour due to their unique characteristics. Thus, assessment, either macro or micro level is important for material development quality and improvement purposes. Micro-inspection approaches such as micrography analysis typically obtained through an image processing instrument (microscope) and a conventional mechanical analysis technique are reviewed. The assessment can ease the complex process and uncertainties during manufacture, assemble and quality control. Substantial it will be able in confirming the manufacturing defect and origin such as variations of volume fractions of fibre and matrix, void in the matrix and between the fibres and matrix, imperfect bonding between constituents and so forth. These studies would particularly be helpful in designing high strength composites and can better withstand the ballistic impact. The uncertainty in these factors circulated to a larger scale and reflected the variability in characterizing the structure performance [7].

This present study aims to fabricate and carry out the preliminary assessment of the aramid fibre reinforced polymer composite material. The composite fabrication method is focused on the hand lay-up lamination method. Each of the composite material specimens is assessed for determining the fabrication related deflect and fibre volume fraction property. The aramid fibre reinforced polymer composite material ballistic specimen development is demonstrated.

2. Material

The composite material is developed based on the composition of matrix material (Epikote 828), a liquid-based epoxy material, and aramid fibre as the reinforcement material.

2.1 Matrix material – Epikote 828

Epikote 828 is a medium viscosity liquid epoxy resin that produced from bisphenol A resin with epichlorohydrin. Cured state Epikote 828 posed a good pigment-wetting, good resistance to the filler setting, and high level of mechanical and chemical resistance properties [8]. The curing agent for Epikote 828 is the modified cyclo-aliphatic amine known as Epikure F205.

Before the lamination process, the matrix solution pot life is determined for allowing a sufficient period of lamination and debulking. The pot life is influenced by the mixing ratio between Epikote 828 and Epikure F205. Series of trials with different mixing ratios had been carried out in determining the pot life 10 minutes, which providing sufficient duration for completing the lamination and debulking process. Based on the trials, a ratio of 50:50 had offered a substantial pot life of 10 minutes compared to other mixing ratios. The mixing ratio provided an adequate mixture viscosity, where it allowed good adhesive flowability and debulking handling in obtaining high bond strength composite.

2.2 Reinforcement material – Aramid fibre

Unidirectional aramid fibre is selected as the polymer composite material reinforcement material. The adopted aramid fibre material (shown in figure 1) is in a roll mat form, plain unidirectional weave fabric, 50 cm width, and density of 280 g/m^2. Based on manufacturer specification, the fibre posed tensile strength of 2200 GPa, and tensile modulus of 110 MPa [9].

Figure 1. Aramid unidirectional plain weave fabric.

3. Material preparation and fabrication
Specimen material preparation and fabrication method is elaborated as follow:

3.1 Preparation of aramid sample plies
Three sets of aramid composite material ballistic test sample are developed, where each sample is laminate with six (06) plies fabric with different plies orientation arrangement as listed in table 1. The plies orientation is arranged for enhancing the composite material properties especially the strength property [1,10]. Detail of the sample plies orientation arrangement is tabulated in table 1 and the composite material orthotropic laminate shown in figure 2.

Table 1. Test sample plies orientation arrangement detail.

Sample	*Plies Orientation*
1	Unidirectional laminate (0°, 0°, 0°, 0°, 0°, 0°)
2	Orthotropic laminate (0°, 90°, 0°, 90°, 0°, 90°)
3	Quasi-isotropic laminate (0°, 45°, 90°, 90°, 135°, 0°)

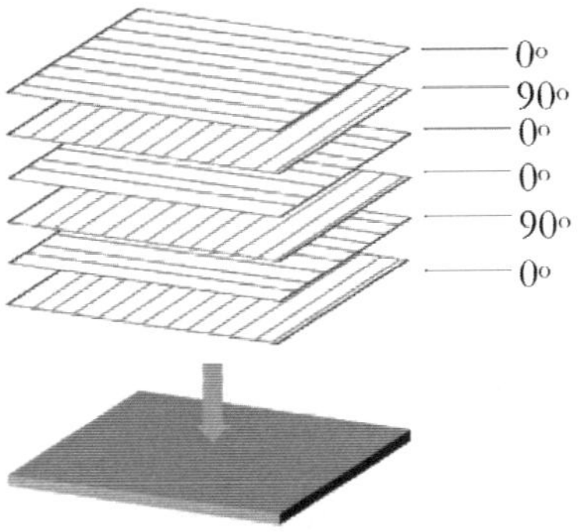

Figure 2. Orthotropic laminate.

Aramid fabric plies prepared according to specific orientation angle by using the transparent template that printed with the desired orientation angle, shown figure 3. The template placed on top of the unidirectional fabric. The cutting reference border is taped with narrow masking tape for ease of cutting.

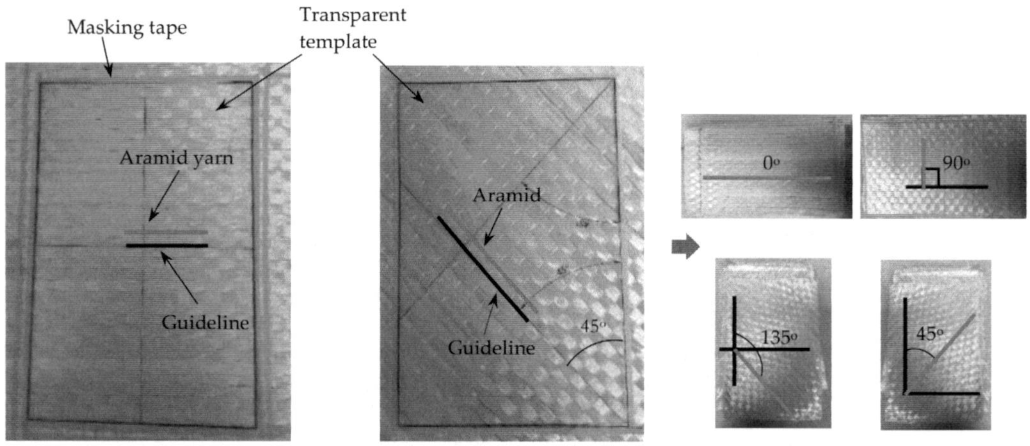

a. 0° and 90° Template b. 45° and 135° Template c. Fibre plies orientation

Figure 3. Aramid fabric preparation procedures.

3.2 Lamination process (Hand lay-up lamination method)

The sample hand lay-up lamination with open mould method shown in figure 4. The top and base plate surface is waxed with mould releaser agent for effective demoulding. The plate surfaces must be clean and clear from the mould releaser agent residual for the lamination process. The matrix solution is prepared at 50:50 (Epikote 828 : Epikure F205), and a small amount of the matrix solution poured and spread around the base plate. The first ply is laid on the solution, and the solution is further distributed to wet the ply thoroughly. L-guide is used to guide the ply to achieve each desired plies orientation, respectively. The process is repeated until the final ply. Then, the top plate lay on the laminate top layer. Additional of 2 kg press load is loaded on top of the top plate to force the excess matrix solution out and determine the sample thickness. The laminate is cured and removed from the mould plate after 48 hours and further rest on the flat surface condition for another 48 hours to ensure fully cured. All the samples followed the same procedures.

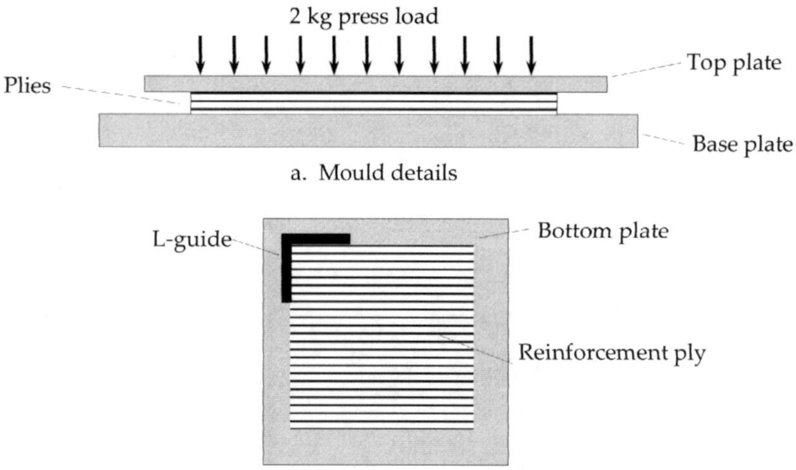

a. Mould details

b. Laminate ply guide by a L-guide.

Figure 4. Hand lay-up lamination method.

IOP Conf. Series: Materials Science and Engineering 1217 (2022) 012008 doi:10.1088/1757-899X/1217/1/012008

4. Physical properties inspection / analysis

Physical properties inspection or analysis is carried out in determining the sample volume fraction property and physical defect due to the fabrication process. The inspection outcomes utilized for improving the fabrication method.

4.1 Fibre volume fraction

The fibre volume fraction is a crucial composite material property, where it defined as the fibre volume percentage in the entire composite material volume. It plays a crucial role in determining the composite material properties through the rule of mixtures. In this study, the sample volume fraction property is determined by measuring the sample physical properties (fibre filament diameter, number of fibre filament per yarn, number of yarn along the section surface and volume of sample). The measurement process is done under micrography observation. All samples (unidirectional, orthotropic, and quasi-isotropic laminate) are fabricated with the same fabrication method and materials. Therefore, the samples have the same volume and thickness property, respectively. The unidirectional laminate sample type is selected in determining the volume fraction property. It is appliable because all three types of laminates have the same density properties. A small block of 25.4 mm length by 12.7 mm width is trimmed from the developed sample. The block longitudinal surface is polished for micrography observation. The micrography observation is carried out by using a digital microscope. Through the micrography observation, the fibre filament diameter and number of fibre filament per yarn is measured, respectively. Figure 5 shown the fibre yarn and filament under micrography observation. The fibre filament average diameter is 13 μm, and the total number of fibre filament of the block is about 36,800 filaments. Therefore, the sample fibre volume fraction is predicted at 0.64.

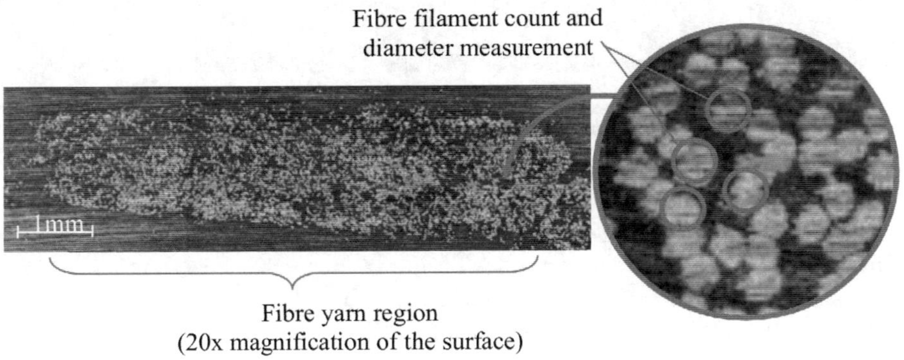

Figure 5. Fibre yarn region and filament.

4.2 Defects

The cured sample specimens are inspected for determining the defects. The inspection is done by physical and micrography observation. Based on the physical observation, there are few visible defects. Defect such as visible gap between the yarns, as shown in figure 6. This defect may due to manufacturing or handling defect, which could cause higher matrix composition and resulting a weak structure strength. Therefore, affected fabric portion is avoided for the sample lamination.

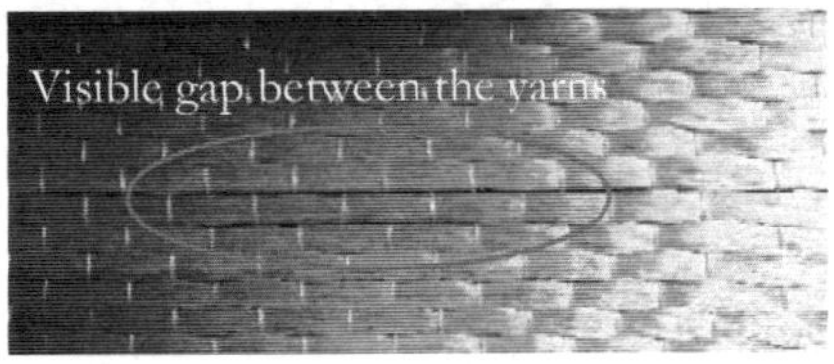

Figure 6. Gap between the yarns.

Physical inspection is carried out on the cured sample for detecting the sample defect, which mainly contribute by the hand lay-up lamination method. There are uneven edge regions shown in figure 7, where it caused by uneven press loading (top mould plate) on the particular lamination corner.

Figure 7. Uneven edge region.

Figure 8 shows that one side of the sample surface was rough, while the other side was smooth. The surface defect might be due to the fabrication process where the debulking process is not sufficient to remove most of the air bubbles. As the top plate mould is placed on the top lamination, air bubbles are trapped in between the top plate surface and form ragged surface as the matrix material is cured.

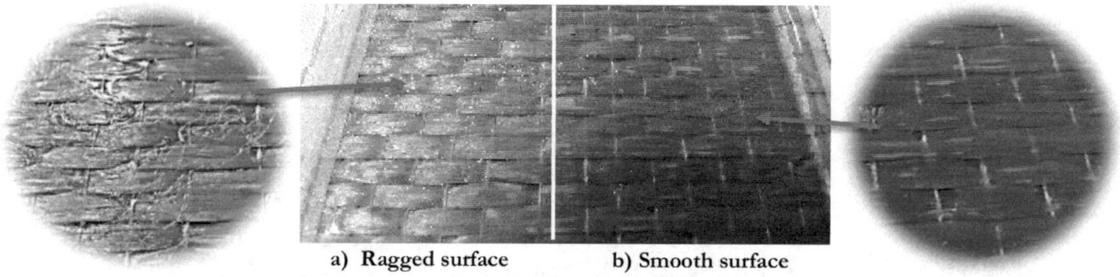

Figure 8. Surface damage.

Based on micrography observation, there is a black spot that displayed an air void from the cross image of composite, shown in figure 9 of a unidirectional laminated composite. The air void may due to the mechanical entrapment of air between plies during fabrication or reaction gases due liquid resin and curing agent chemical reaction. Hence, excessive porosity in laminate leads to reduced mechanical properties, notably reduced interlaminar shear and compression strength [11].

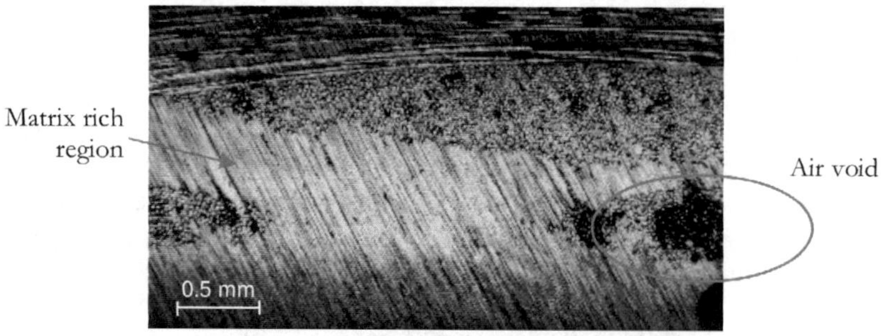

Figure 9. Air bubble visible (5x magnification of the surface).

5. Conclusions

Aramid fibre reinforced polymer composite material had been developed, where the polymer matrix is formulated based on Epikote 828 material. Three types of laminates (unidirectional, orthotropic and quasi-isotropic cross plies) were developed. Effective orientation laminate enhanced by utilized the transparent template in preparing the desired aramid fabric orientation. The sample laminated by hand lay-up method with two pieces moulding method. Each laminate laid and guided by an L-guide to ensure all the plies are aligned to each other. Quality assessments include of fibre volume fraction prediction and defect identification are carried out to assess the developed sample materials quality and defect. The assessment outcomes are feed in to improve of the material formulation and fabrication method.

Acknowledgements

This study had been carried out by the researchers of Universiti Malaysia Sabah and funded by Universiti Malaysia Sabah (GUG0337-2019).

References

[1] Elanchezhian C, Abhijith M, Vijaya R B, Rajesh S, Kathir K K and Dinesh R R 2018 *Materials Today: Proc.* **5** 1156–61
[2] Gong R and Chen X 2015 *Technical Yarns* (Handbook of Technical Textiles 2nd Ed vol 1) ed Horrocks A R and Anand S C (Sawston: Woodhead Publishing Limited) chapter 3 pp 43–62
[3] Sharma S, Rawal J, Dhakate S and Singh B P 2020 *Composites Sci. and Technol* **199** 108370
[4] Mallick P 2014 *Fiber- Reinforced Composites* 3rd Ed (London: CRC Press)
[5] Powell P C 2012 *Engineering with fibre-polymer laminates* (Berlin: Springer Science & Business Media)
[6] Jamir M R M, Majid M S A and Khasri A 2018 *Sustainable Composites for Aerospace App* 155–70
[7] Shaw A, Sriramula S, Gosling P D and Chryssanthopoulos M K 2010 *Composites Part B: Eng* **41**(6) 446-53
[8] Yilmaz H, Bedir F and Elmas U 2014 *Proc. The Int. Conf. on Applied Mechanics and Mechanical Eng.* **16**
[9] TEIJIN 2012 Twaron – a versatile high-performance fiber *Twaron Product Catalog.*
[10] Qiao P, Yang M and Bobaru F 2008 *J. of Aerospace Eng.* **21**(4) 235–48
[11] Wood J R and Bader M G 1994 *Composites Manufacturing* **5**(3) 139-47

Feasibility study: Resin-based functionally graded material incorporated with carbonized waste rice husk

G J H Melvin[1*], H M Ling[2], W M J Karen[2]

[1] Centre of Research in Energy and Advanced Materials, Faculty of Engineering, Universiti Malaysia Sabah, Jalan UMS, 88400 Kota Kinabalu, Sabah, Malaysia.
[2] Faculty of Engineering, Universiti Malaysia Sabah, Jalan UMS, 88400 Kota Kinabalu, Sabah, Malaysia.

*Corresponding Author: melvin.gan@ums.edu.my

Abstract. Feasibility study was conducted in exploring the fabrication and characterization of resin-based functionally graded material (FGM) incorporated with carbonized waste rice husk. The waste rice husks were converted into carbon materials through heat treatment under the presence of inert gas at 500°C for 2 hours. Then, they were incorporated into resin to form FGM by centrifugal method to achieve desired gradation. Sample B3 with 5 wt.% of carbonized rice husk (CRH) incorporated into polyester resin (including hardener and ethanol) was centrifugated at 4000 rpm for 30 minutes to form FGM. The fabricated samples were cut into three parts, namely upper, middle, and bottom layer to further characterize the properties at various gradation levels. The density of sample B3 increased gradually, 4.10%, 6.54%, and 6.93% when compared to bulk resin, from upper to bottom layer, respectively. The hardness of sample B3 increased gradually, 27.38%, 42.57%, and 47.08% in contrast to bulk resin, from upper to bottom layer, respectively. FGM proposed in this study can be further manipulated based on the centrifugal force and time, ratio of solvents/hardener, and weight percentage of CRH that indicate they can be exploited for specific of numerous appropriate applications.

1. Introduction

Functionally graded material (FGM) is considered as a type of composite material, where the composition and structure vary gradually over volume, resulting in distinguishable characteristics corresponding to the gradation or heterogeneity [1]. The concept was developed in Japan in 1984, in order to achieve thermal gradient across material's cross section. The continuously changing properties and functions of FGM also provide beneficial advantages, such as to avoid material's delamination/crack, to avoid stress concentration in material, and others [2]. Since then, the concept is further established with various improvements for numerous applications, such as in aerospace, electronics, energy conversion field, and so on.

Carbon materials such as carbon nanotubes, vapor grown carbon fibre, and their derivatives/hybrids, are widely utilized as the filler materials in composite materials for reinforcement/functionality, including FGM [3-6]. Alternatively, carbon materials derived from agricultural waste materials such as rice husk (RH) or sawdust, offer a unique opportunity, due to the low cost and abundancy [7-10]. Commonly, RH is exploited as low value energy resource, discarded, or simply burnt at the field, which will influence the air quality and unfavourable to the environment. Thus, the utilization of carbonized RH (CRH) can unlock its potential as filler materials or for comprehensive applications.

Content from this work may be used under the terms of the Creative Commons Attribution 3.0 licence. Any further distribution of this work must maintain attribution to the author(s) and the title of the work, journal citation and DOI.
Published under licence by IOP Publishing Ltd

Fabrication of FGM can be majorly categorized into gas-based method, liquid phase process, and solid phase method [11]. Comparatively, centrifugal method in liquid phase process can be considered as the easy and low-cost route. In this feasibility study, the ability for carbonized RHs to form gradation inside resin by using centrifugal method was investigated. The gradations were further confirmed through observation, density comparison, and hardness test.

2. Experimental

Dried RHs without further modification were used as the carbonization source. They were carbonized at 500°C, 2 h using furnace, under the presence of argon gas. Then, they were cooled down naturally, crushed into powder and sieved to 90-180 µm.

FGM were formed by loading the CRHs into the mixture of polyester resin, methyl ethyl ketone peroxide (MEKP, hardener), and ethanol; and then the mixture was centrifugated at 4000 rpm for 0.5 h. The ethanol:resin ratio is 3:10, and 5 samples were fabricated as tabulated in table 1. Sample B1 with 1 wt.% CRH loading was prepared without centrifugation as a control sample, and sample B2-B5 was prepared by using centrifugal method with different CRH loadings varied from 1 – 20 wt.%. Ethanol was utilized to dilute the mixture to appropriate viscosity for FGM fabrication, and MEKP:resin ratio is 1:50 which allows sufficient solidification time after the mixture is centrifugated.

Table 1. Fabricated FGM samples.

Sample	CRH contents (wt.%)
B1 (without centrifugal)	1
B2	1
B3	5
B4	15
B5	20

The fabricated samples were cut into three parts with diameter and height about 2 cm and 4 cm, respectively. The parts are denoted as upper, middle, and bottom layer to further characterize the properties at various gradation levels. Suspension method was utilized to determine the density for each layer. Hardness measurement was performed by using Rockwell hardness tester (HRF, 60 kgf) for each layer. The samples were observed using scanning electron microscopy (SEM; Hitachi S-3400N), in order to investigate the structure and morphology.

3. Results and discussions

The images related to CRH and FGM layers are as shown in figure 1. From figure 1 (a), the CRH is black in appearance, and this resulted in the FGM is unable to be distinguished based on colour. Thus, the gradation evaluation based on density and hardness was conducted. SEM image of crushed and sieved CRH is shown in figure 1 (b). FGM, sample B3's respective layer is shown in figure 1 (c).

Figure 1. Image of (a) CRH, (b) SEM: crushed CRH and (c) FGM.

IOP Conf. Series: Materials Science and Engineering 1217 (2022) 012009 doi:10.1088/1757-899X/1217/1/012009

The density of bulk resin was 1.1527 g/cm^3, within the range of typical polyester resin [12]. The density increment pattern of fabricated FGM when compared to bulk resin, is illustrated in figure 2 (a). Except for sample B1, the density for all samples increased gradually from upper to bottom layer. Furthermore, the density of sample B5 also increased with the highest value, as the CRH content increased. The density of each layer increased gradually from upper layer to bottom layer according to the direction of centrifugal force. The increasing density trend also had been observed for different FGM incorporated with nano filler materials, and fabricated through centrifugal method [4, 13]. For sample B1, it can be considered as no appropriate gradation occur, as centrifugal force is not applied. The gradation of sample B1 is highly dependent on resin's viscosity and CRH's movement within solidification time, in which longer time might be required for CRH to shift from upper to bottom layer without centrifugal force.

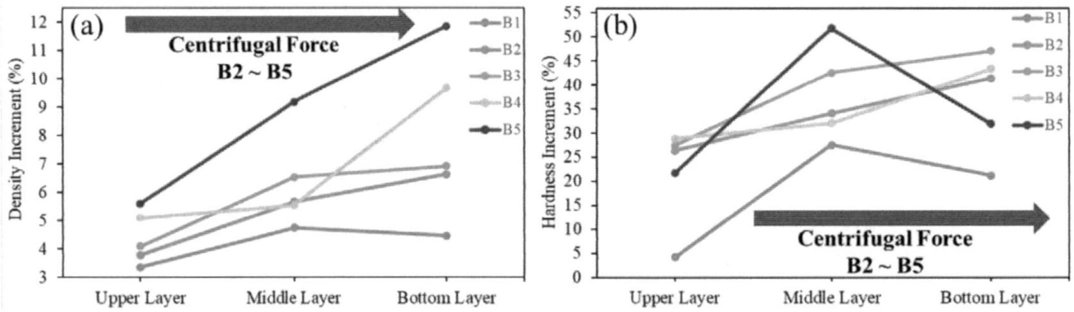

Figure 2. Increment percentage (a) density and (b) hardness.

The hardness of bulk resin was 59.9 HRF, within the range of typical polyester resin [14]. The hardness increment pattern of fabricated FGM when compared to bulk resin, is depicted in figure 2 (b). The hardness of sample B1 and B5 increased from upper to middle layer but decreased at bottom layer. For sample B1, it can be associated with the density's pattern, where the increment of hardness can be related to the percentage of CRH contents (B1 bottom layer's density decreased). Apparently for sample B5, there is a limit to CRH bonding with the resin, where at certain saturated level produces an adverse result on hardness. Conversely, the hardness increased gradually from upper to bottom layer for sample B2, B3, and B4, respectively. Comparatively, sample B3 exhibited the highest hardness increment, 27.38% to 47.08% from upper to bottom layer. The increasing hardness trend as the filler contents increased, also can been observed for various composite materials including FGM [13, 14].

Based on the results stated above, sample B2, B3, and B4 indicated that they can be considered as the successful FGM with appropriate gradation on density and hardness. Among them, sample B3 showed gradually increasing density and most significant hardness increment, from upper to bottom layer.

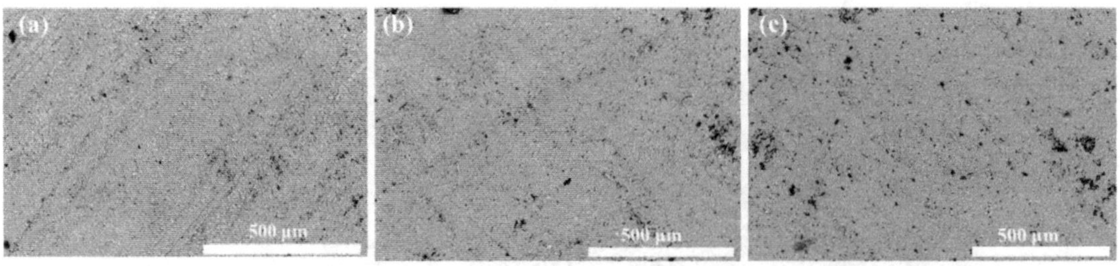

Figure 3. SEM image of sample B3 (a) upper, (b) middle and (c) bottom layer.

The SEM cross section images of sample B3 for each layer is shown in figure 3. Other than porosity, there were no other significant discontinuities in the structure. The amount of pores gradually increased, from upper to bottom layer. With higher contents of CRH at bottom layer, based on their density, higher number of pores are introduced. Furthermore, the centrifugal force also shifted the CRH particles to bottom layer, which contributed to the increase in pore numbers. This pattern is in accordance with Ref. [15], and further indicates the successful fabrication of FGM.

4. Conclusions

Carbon materials derived from RHs were prepared by carbonizing them at 500°C for 2 hours in inert atmosphere. CRHs were incorporated into polyester resin with various loadings to form FGM by centrifugal method at 4000 rpm for 30 minutes, to achieve the desired gradation. Sample B3's density increased gradually, 4.10%, 6.54%, and 6.93% when compared to bulk resin, from upper to bottom layer, respectively. Furthermore, sample B3's hardness increased gradually, 27.38%, 42.57%, and 47.08% in contrast to bulk resin, from upper to bottom layer, respectively. As the CRH contents increased from upper to bottom layer for sample B3, number of pores also increased which further indicated the successful fabrication of FGM. Based on the results obtained, it is feasible to fabricate FGM by incorporating CRHs into polyester resin. Furthermore, with appropriate centrifugal force and time, ratio of solvents/hardener, and weight percentage of CRH, different variations of FGM can be exploited.

Acknowledgements

This work was supported by Universiti Malaysia Sabah (Grant number GUG0386-2/2019), Malaysia.

References

[1] Miyamoto Y, Kaysser W A, Rabin B H, Kawasaki A and Ford R G 1999 *Functionally Graded Materials: Design, Processing and Applications* (Boston: Springer)

[2] Bhavar V, Kattire P, Thakare S, Patil S and Singh R K P 2017 *IOP Conf. Ser.: Mater. Sci. Eng.* 229 012021

[3] Wang Z, Jin K W M and Melvin G J H 2020 Carbon materials from various sources for composite materials *Composite Materials: Applications in Engineering, Biomedicine and Food Science* ed S Shafiquzzaman, G J H Melvin and R Mizanur (Cham: Springer) chapter 1 pp 3-33

[4] Wang Y, Ni Q-Q, Zhu Y and Natsuki T 2014 *Nano* **9**(1) 1450011

[5] Wang Y, Ni Q-Q, Zhu Y and Natsuki T 2013 *Polym. Compos.* **34**(10) 1774-81

[6] Melvin G J H, Ni Q-Q and Wang Z 2017 *Phys. Status Solidi A* **214**(2) 1600541

[7] Karen W M J, Ooi JX, Wang Z, Siambun N J, Liew W Y H, Muzammil W K, Sopian K and Melvin G J H 2021 *Int. J. Surf. Sci. Eng.* **15**(1) 36-49

[8] Melvin G J H, Chai K F and Tamiri F M 2019 *IOP Conf. Ser.: Mater. Sci. Eng.* 606 012002

[9] Melvin G J H, Wang Z, Siambun N J and Rahman M M 2017 *IOP Conf. Ser.: Mater. Sci. Eng.* **217** 012017

[10] Karen WM J, Wong J H, Wang Z, Siambun N J, Liew W Y H, Muzammil W K, Sopian K and Melvin G J H 2021 *Mater. Chem. Phys.* **263** 124371

[11] Naebe M and Shirvanimoghaddam K 2016 *Appl. Mater. Today* **5** 223-45

[12] Alhumdany A A, Al-Waily M and Kadhim M H 2015 *Int. J. Energy Environ.* **6**(5) 499-516

[13] Wang Y, Ni Q-Q, Zhu Y and Natsuki T 2014 *Polym. Compos.* **35**(3) 557-63

[14] Anand P, Padmaraj N H, Chethan K N and Vibhor S 2017 *Mater. Res. Innovations* **21**(2) 115-21

[15] Strojny-Nędza A, Pietrzak K and Węglewski W 2016 *J. Mater. Eng. Perform.* **25**(8) 3173-84

IOP Conf. Series: Materials Science and Engineering 1217 (2022) 012010 doi:10.1088/1757-899X/1217/1/012010

A review on the development of electro-carburisation process

S C On[1*], S A Bahrin[3], M Gakim[1], W Y H Liew[1], F M Tamiri[1] and N J Siambun[1,2*]

[1] Mechanical Engineering, Faculty of Engineering, Universiti Malaysia Sabah, Jalan UMS 88400 Kota Kinabalu, Sabah, Malaysia.
[2] Centre of Research in Energy and Advanced Materials, Faculty of Engineering, Universiti Malaysia Sabah, Jalan UMS 88400 Kota Kinabalu, Sabah, Malaysia.
[3] Department of Mechanical Engineering, Polytechnic Kota Kinabalu, Jalan Polytechnic, 88450 Kota Kinabalu, Sabah, Malaysia

[*]Corresponding author: nancyjs@ums.edu.my, BK16110206@student.ums.edu.my

Abstract. The purpose of this paper is to review the early development of electro-carburisation technology and the research findings related to the electro-carburisation process. In general, conventional liquid carburisation of steel using a molten cyanide bath is carried out to improve the performance of mild steel, however this process produces toxic cyanide waste. Thus, other alternatives for liquid carburisation are necessary. Electro-carburisation process using carbonate-base molten salt, under a CO_2 environment was developed as one of the alternatives to liquid carburising. Metal to be treated is exposed to the carbon-rich liquid in the molten cyanide bath and electro-carburisation. However, the metal is simply immersed inside the cyanide bath during conventional liquid carburising, while connected to the cathode in the electro-carburisation. The electro-carburisation involves a diffusion of carbon atoms into the surface of the metal which enhance the surface hardness of the metal. The effects of electrolysis parameters to the surface hardness and case hardening of treated metal have been reported in several journals. This article summarises the research findings. Apart from that, the quenching process and heat treatment post quenching also plays an important role in the quality of the carburised metal, therefore also reviewed in this article.

1. Introduction

Carburisation is a common practise to increase the carbon content on the surface of metal to obtain a harder martensite phase upon quenching process. This produces extra hardness and resistance to wear and contact fatigue. The metal's core however remains unchanged thus relatively softer and able to withstand impact [1]. The introduction of carbon atoms into the surface of metal can either be in the form of solid, gas, vacuum, plasma or liquid [2]. In this article, introduction of carbon in liquid form is of interest. Basically, the metal to be treated is placed inside a liquid medium, often known as a salt bath or molten salt, which contains a rich carbon source. It is reported that the liquid-carburising has a short cycle time because the heating is faster in the molten salt solution [3]. The obtained case hardening is comparable to the gas-carburising in some ammonia environment. In liquid carburising, the carbon source is either from (i) decomposition of cyanide, (ii) addition of carbon source into the molten mixture,

Content from this work may be used under the terms of the Creative Commons Attribution 3.0 licence. Any further distribution of this work must maintain attribution to the author(s) and the title of the work, journal citation and DOI.
Published under licence by IOP Publishing Ltd

or (iii) reaction occurred in the molten salt to produce carbon. Thus, a cyanide or non-cyanide molten salt could be used in the liquid carburising process [3].

1.1 Carbon source generated from chemical reaction in cyanides bath

Molten salt baths containing cyanides have been used for many years for the surface treatment of steel. One early-stage molten salt carburisation used cyanide salt bath $Ba(CN)_2$. The liquid carburising process was carried out at 870°C for 10 hours. During this process, the $Ba(CN)_2$ decomposed to carbon according to the reaction 1 and allowing the carbon diffused into the surface of steel led to carburisation process [4].

$$Ba(CN)_2\ (l) \rightarrow BaCN_2\ (l) + C\ (s) \tag{1}$$

The formation of a cyanide bath could be prepared from a mixture of cyanates of lithium, sodium or potassium, with an addition of a very small amount of sulphur compounds and carbonates. At the operating temperature of 565 ± 5°C, the cyanate forms carbonates as shown in reactions 2 and 3. The carbon monoxide then breaks to carbon according to reaction 4. The carbon is then diffused into the metal's surface thus increasing the carbon content inside the iron lattice, causing surface hardness [5].

$$4CNO^- \rightarrow CO_3^{2-} + CO + 2N + 2CN^- \tag{2}$$

$$2CNO^- + O_2 \rightarrow CO_3^{2-} + CO + 2N \tag{3}$$

$$2CO \rightarrow CO_2 + C \tag{4}$$

In some salt bath mixtures, an activator is necessary to be added to replenish the material and to ensure the bath is workable for long period of time [6]. The operating temperature and period of liquid carburising containing cyanide can either be carried out in a low temperature (845 – 900°C) to obtain light case (0.13 – 0.25 mm), or at high a temperature (900 – 955°C) for deep cases (1.0 – 2.0 mm) [3]. The environmental problems associated with disposal and handling of the toxic cyanide salt baths, is no longer acceptable due to the tightening of anti-pollution measures by the authorities. This has restricted the use of this process therefore eliminating the toxic hazard of cyanide is required. One reported study [7] has tried to reduce the formation of irritating and poisonous gasses released from the process, through burning the gas at the surface of the melt to produce a non-poisonous and non-irritating gas.

1.2 Carbon source generated from chemical reaction in non-cyanide bath

The need to eliminate the usage of a toxic cyanide salt bath in the liquid carburising process has encouraged several researchers to replace the salt bath to a non-toxic salt bath. Usage of non-toxic compounds containing alkali metal chloride, alkali metal carbonate and replenishment materials such as graphite [8-9] or silicon carbide [10] have been used. A propeller stirrer may be used to ensure uniform dispersion of the replenishing materials.

In the research of Kerridge in 1984, a mixture of sodium carbonate (Na_2CO_3), sodium chloride (NaCl) and silicon carbide (SiC) was used [10]. The NaCl was added to lower the melting temperature of the carbonate salts. Kerridge, claimed that the carburising species were generated by a reaction between carbonate ions and silicon carbide to form CO according to reaction 5 [10].

$$3Fe\ (s) + SiC\ (s) + 2Na_2CO_3\ (l) \rightarrow Na_4SiO_4\ (l) + Fe_3C\ (s) + 2CO\ (g) \tag{5}$$

The disadvantage of this method was the formation of a thick silicate crust on the melt surface, making the salt bath unworkable. The use of chloride increases the hardness of the crust that makes it difficult to break when cooled. Consequently, the addition of costly Li_2CO_3 was required. It appears that the lithium ions can prevent encrustation of the melt surface by encouraging the silicate to crystallise out as needles upon cooling. Such silicate could easily be removed from the molten bath by sieving. At molten salt temperatures between 900 – 955°C the obtained case hardening is 0.5 – 3.0 mm, similar to the high temperature bath containing cyanide. Studies show that, heat treated metal in a non-cyanide bath produced less retained austenite upon quenching process, compared to those using cyanide bath [3].

1.3 Carbon source generated from electrolytic process

In the liquid carburising process, the carbon source could also generate via electrolytic process. Two patents were filed by Thomas J. Buitkus (1967) and Hiyama Shinji (1975). In the Thomas invention, the carburisation process was affected by the electrolytic decomposition of molten cyanogen compounds, which resulted in a uniform distribution of carbon in reactive contact with the articles being carburised. The metal to be carburised was made the anode and a porous electrode the cathode. The author stated that this process provides a continuously generated carbon source for the carburisation process and leaves no insoluble coating on carburised material after quenching [11]. Hiyama used a salt mixture of Na_2CO_3-NaCl (0.25:0.75 mole ratio), with the metal to be carburised serving as the cathode and inert graphite as the anode. During the electrolysis process, carbonate salts decomposed to form carbon entities, which then diffused into the metal structure and led to carburisation [12]. Hiyama claimed that this process was easier to control and no sludge was formed as well as permitting prolonged use of the salt bath. This process is a promising method to eliminate the usage of a toxic cyanide bath and formation of a thick crust on the melt surface [12].

After decades of being patented, attempts to duplicate Hiyama's process in the laboratory has found that the graphite used as anode can reduce in size. It was found that the graphite experiences erosion due to the anodic reaction, in which the oxygen ion reacts electrochemically with carbon to produce CO_2 gas [13]. Popov *et al.* reported that anode materials such as graphite are often easily consumed by anodic reactions involved in high temperature molten salts [14]. It was concluded that graphite was not suitable to be used as an anode. It was also found that the depth of the case hardening of treated metals were not consistent after salt is repeatedly used [13]. Thus, intensive investigations have been carried out to further understand and improve the process. This review aims to discuss the findings of the investigation and improvement to the liquid carburisation of Hiyama method via electrolysis process. Besides that, the effects of post heat treatment such as quenching is also covered. The studied and improved liquid carburising via electrolysis process is termed as the molten salt electro-carburisation process.

2. Carburisation process and case hardening determination

Generally, during the carburisation process, carbon atoms diffused into the crystal lattice of the low carbon steel causing to lattice strain, thus increase the metal hardness and strength. The depth of carbon diffusion is also known as case depth. If quenching is carried out right after the carburising process it leads to the transformation of austenite to martensite and appears as retained austenite at room temperature [20]. The hardenability and type of quench are determined by section size, desired core hardness, and service requirements [15].

The reported studies of electro-carburisation processes, factors such as carbon availability in the molten salt, carburising time, as well as the quenching processes have been investigated [2, 13, 18]. The surface hardness and case hardening analysis was carried out using a hardness testing machine, for example Rockwell, Knoop, Brinell and Vickers Hardness. In determining the depth of carburisation, the carburised metal has to be cut cross-sectionally exposing an area not exposed to the molten salt (figure 1a & b), then the indenter was pressed on the cross-section area from surface towards the core.

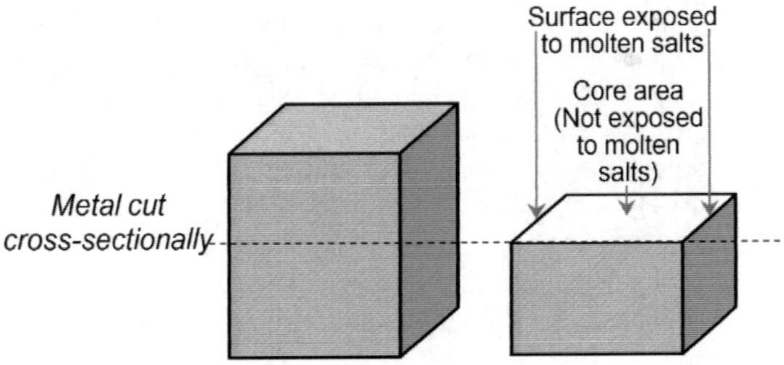

Figure 1. Electro-carburised rectangular shape metal (a) before being cut, and (b) after cut cross-sectionally to exposed the core and area non-exposed to molten salt. Surfaces exposed to molten salt is shaded grey areas.

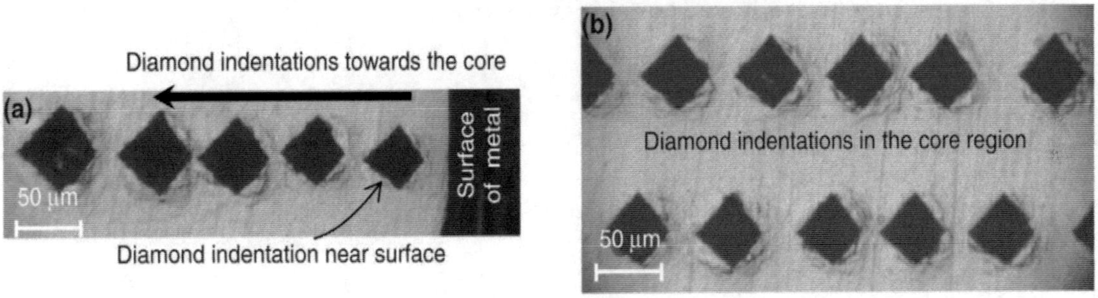

Figure 2. Indentations recorded in Vickers micro-hardness tests on an electro- carburised steel sample (a) from the surface towards the core, and (b) in the core region [2].

3. Molten salt electro-carburisation process

The liquid carburisation process via electrolysis process was carried out in a sealed stainless steel tube reactor with continuous purged of CO_2. A crucible containing salt was placed inside the centre of tube reactor to obtain optimum heat. The metal that needed to be carburised was pre-connected to the cathode. Once the salt melted, a K-type thermocouple, anode and cathode/metal to be carburised were immersed completely inside the molten salt. The reactor was then sealed securely, ready for electro-carburisation process. The setup of molten salt electro-carburisation process is shown in figure 3.

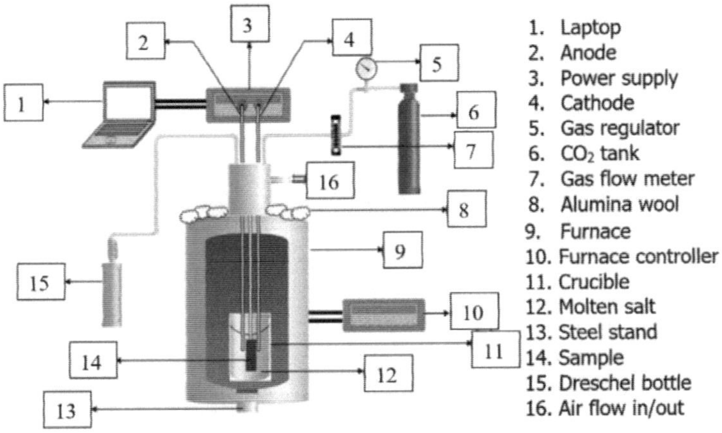

Figure 3. The schematic setup of electro-carburisation.

The carburisation process is initiated by applying voltage through the pair of electrodes. The carbonate ions (CO_3^{2-}) decompose to C and O^{2-} at the cathode according to reaction 6. The nascent carbon atoms then diffuse into the metal to be carburised at the cathode. The O^{2-} ions remain in the molten salts to either react with the CO_2 at the molten salts/gas interface to reform the CO_3^{2-} ion according to reaction 7, or transport to, and discharge at the inert anode to produce O_2 gas according to reaction 8. The overall reaction of the electro-carburisation process is the extraction of carbon from CO_2 into iron with O_2 gas as the by-product as shown in reaction 9.

$$CO_3^{2-} + Fe + 4e \rightleftharpoons C\text{-}Fe + 3O^{2-} \quad (6)$$

$$O^{2-} + CO_2(g) \rightleftharpoons CO_3^{2-} \quad (7)$$

$$2O^{2-} \rightleftharpoons O_2\,(g) + 4e \quad (8)$$

$$CO_2 + Fe \rightleftharpoons C\text{-}Fe + O_2\,(g) \quad (9)$$

4. Electro-carburisation process parameters

Intensive investigation has been carried out to understand the electro-carburisation process by researchers. Four electro-carburisation parameters were identified to directly give effects to the surface hardness and the case-hardening of treated metal. The electro-carburisation parameters are CO_2 gas environment, type of salts as electrolytes, period of carburisation and the electrolysis voltages. The findings are further discussed in the following.

4.1 CO_2 gas environment

In the electro-carburisation process, the source of carbon for the carburisation process is from decomposition of carbonate ions into carbon as indicated by reaction 6. The metal oxide that was formed simultaneously in reaction 6, react with CO_2 gas to form a carbonate compound according to reaction 7, thus ensuring continuous availability of a carbonate compound in the molten salt.

Studies have been carried out to determine the effect of CO_2 gas in the electro-carburisation process. In one investigation by Siambun *et al.*, the gas flowing into the system was maintained at 200mL.min^{-1}, however the CO_2 concentration was reduced by mixing the gas with N_2 gas. It was shown that the higher the concentration of CO_2 gas in the electro-carburisation system, the higher the availability of carbonate ions and carbon for the carburisation process. The wt.% of carbon derived from the recorded electrolysis total charge has shown a reduction in carbon availability when the CO_2 gas concentration was reduced. Consequently, the obtained surface hardness and case hardening were lower. The higher CO_2 partial pressure tend to increase the hardness of the sample as the case hardening is caused by the diffusion of carbon atoms into the crystal lattice of iron during the electro-carburisation process. This causes the internal straining and thus increase the hardness of metal [16].

4.2 Type of salts as electrolytes

It is important that the salt mixture has high thermal stability at the carburising temperature (800°C or higher), but at the same time the melting temperature of the salt mixture needs to be appropriately low for low thermal energy consumption, but sufficiently high so that the electro-reduction of the CO_3^{2-} ion to solid carbon is easier in terms of both kinetics and thermodynamics.

In the investigation reported by Siambun *et al.* [2], two salt mixtures have been studied i.e. pure carbonate salt and carbonate-chloride mixture. The salt mixture was formulated by taking into consideration the working temperature, thermal stability, and rate of electro-carburisation (reduction of CO_3^{2-} to form C). Li_2CO_3, K_2CO_3 and Na_2CO_3 are thermally stable and thermodynamically reactive thus easy to undergoes reaction 1. The melting temperature of Li_2CO_3 (720°C) is lower than 800°C, however usage of a single salt alone is quite costly. The authors used Li_2CO_3-K_2CO_3 (mole ratio 1:1) to obtained liquidus phase ~503°C. The melting temperature of Na_2CO_3 (~858°C) is higher than 800°C. Mixture of Na_2CO_3-NaCl (Mole ratio 4:1) reduce the melting temperature to ~740°C. Electro-carburisation process in both salt mixtures were carried out by applying a constant 2.5 V to a pair of electrodes; inert anode (10 mm diameter ELR SnO_3 rod 98.5% purity) and low carbon steel (0.1~0.2 wt.%C; diameter = 5 mm) sample as cathode, for 1 hour at 800°C salt temperature and in a CO_2 gas environment. Samples electro-carburised in pure carbonate and carbonate-chloride mixture shows dominance of the cementite phase in the sample with almost no sign of isolated ferrite phase under the backscattered electrons (BSE) analysis. The cementite phase dominance in the pure carbonate salt however could be obtained at a lower voltage, suggesting availability of carbon in pure-carbonate is higher than the carbonate-chloride mixtures. This is also evident from the wt.% of carbon derived from the recorded electrolysis total charge value. Under 2.4 V, Na_2CO_3-NaCl produced 9.13 wt.%C, while Li_2CO_3-K_2CO_3 produced 12.17 wt.%C. Analysis using Vicker's micro-hardness obtained maximum average surface hardness of 1175 ± 25 HV and a case-hardening depth of 0.45 – 0.47 mm [2]. Study of wear on electro-carburised metal was reported to reduce the severity of adhesive wear and hence the tendency of the worn surface to fracture [17].

It is important to note that due to the allowable carbon solubility into the lattice of mild steel at 800°C, some of the carbon may only present on the surface of the cathode. Besides that due to the reactivity of Li_2CO_3-K_2CO_3, applying a higher voltage would risk formation of carbon on the surface of metal to be carburised, that resulted retardation of the carburisation process [13].

4.3 Period of carburisation
During the electro-carburisation process, when the surface of the metal is exposed to a carbon rich molten salt, the carbon is immediately absorbed into the iron lattice. If the carbon potential inside the molten salts is high enough and supplied consistently, the metal surface reaches its carbon saturation [13]. For instance, electro-carburisation carried out in lower CO_2 gas concentration (Flow rate of 100 $mL.min^{-1}$ CO_2 + 100 $mL.min^{-1}$ N_2) is still able to reach its carbon saturation after being exposed for a long period of time, provided the carbon continuously has sufficient driving force.

In one study by Liew *et al.*, electro-carburisation in 860°C Na_2CO_3-NaCl molten salt was carried out to two specimens (75 x 10 x 4 mm mild steel) under 200 $mL.min^{-1}$ CO_2 gas flow rate and 4.5 V. One specimen underwent 1-hour electrolysis process, and another one for 3-hours. Under these process conditions, the carbon concentration was consistent throughout the specified length of process. However due to longer period of exposure to the carbon source, the 3-hour process obtained higher surface hardness and case depth (795 HV & 450 μm), compared to the 1-hour process (727 HV & 50-100 μm) [18].

4.4 Electrolysis voltages
During the electro-carburisation process, voltage is applied to allow reactions 6 to 9 occurring. The availability of carbon atoms depends on the voltage applied during the electro-carburisation. A study to investigate the effect of voltage variation reported by Siambun *et al.* [2]. Electro-carburisation processes were carried out on a 5mm diameter mild steel specimen (0.1~0.2 wt.%C), at 800°C Na_2CO_3-NaCl molten salt in CO_2 environment for 1 hours, but at various voltages (1.0V to 2.5 V). It was reported that the charge value was higher when a higher voltage was applied, which indicated rapid reactions 3 to 4 activities. The surface hardness and depth of case-hardening values were reported to be highest for the 2.5 V process, followed by 2.0 V, 1.5 V and 1.0 V. With enough duration of exposure, the same surface hardness and depth of case-hardening may be obtained for processes at a lower voltage [2].

5. Post heat treatment
In the electro-carburisation process, treated metal specimens were commonly quenched to lock the carbon that had diffused into the iron lattice. A study by Siambun *et al.* has shown a significant difference when the treated sample cooled naturally in air and quenched in water. When the carburised

sample cooled at slow rate, the carbon atoms are able to migrate to separate layers in the microstructure and form a pearlite which is softer [19].

5.1. Quenching medium

In the study reported by Bahrin *et al.*, metal specimens sizing 25 x 12 x 20 mm (0.1 ~ 0.15 wt.% C) was electro-carburised in 850 ±10°C Na_2CO_3-NaCl molten salt in CO_2 environment for 3 hours and under 4.0 V, and thereafter were rapidly cooled in two quenching medium i.e. water and oil. The study found that specimens quenched in water has a higher surface hardness and depth of case-hardening (1440 ± 5 HV; 800 μm), compared to the oil quenched (1005 ± 5 HV; 400 μm). Specimens quenched in oil shows a significant dip in hardness near the surface. Microstructure analysis confirmed the presence of austenite trapped between the plate-like structure that contributes to the low hardness [20]. As reported by Torkamani *et al.*, high viscosity quenching medium is less efficient in heat removal during quenching. Future study to determine the effect of others type of quenching medium used after the electro-carburisation process can be carry out [21].

5.2. Post heat treatment process

One of the limitations that has been found in the electro-carburisation process is the formation of retained austenite at the near-surface upon quenching [18, 20],. The retained austenite decreased the hardness, and when undergoing grinding will produce cracks [22]. One suggested method to tackle this problem is to reheat the carburised part, to allow diffusion of carbon thus moderating the carbon content at the surface. Upon quenching, the amount of retained austenite becomes less at the surface. Generally, less than 15% of retained austenite can be tolerated depending on its application [22]. Reheating the carburised part is known as the post heat treatment process. Perhaps further investigation on the post heat treatment should be carried out to improve the electro-carburised metal quality. Other treatment (normalizing, annealing, tempering etc.) that is able to modify the microstructure and mechanical properties of the metals are worth investigating.

6. Conclusion

The molten salt electro-carburisation is a promising method to be used as an alternative to the conventional toxic cyanide bath liquid carburisation. This process uses a molten carbonate-base as a medium to dissolved CO_2 gas and ensure continuous supply of carbon for the carburisation process. Careful selection of the salt mixture is important to ensure the thermal and chemical stability during the electrolysis process. While the molten salt temperature determines the amount of carbon that can be dissolved in the iron lattice, the surface hardness and depth of case-hardening could be controlled via electrolysis voltage and the period of exposure. Post electro-carburisation process is equally important. Proper quenching and post treatment increases the quality of the carburised specimen. Nevertheless, many studies need to be explored to improve the electro-carburisation process.

Acknowledgement

This work was supported by the Ministry of Higher Education, Malaysia (Fundamental Research Grant Scheme FRGS0470/2017)

References

[1] Lingamanaik S N and Chen B K 2012 The effects of carburising and quenching process on the formation of residual stresses in automotive gears *Computational Mat. Sci.* **62** 99– 104
[2] Siambun N J, Mohamed H, Hu D, Jewell D, Beng Y K and Chen G Z 2011 Utilisation of carbon dioxide for electro-carburisation of mild steel in molten carbonate salts *J. Electrochem. Soc.* **158**(11) H1117
[3] Chandler H 1995 *Heat Treating Processes and Related Technology Introduction* (International A.S.M)
[4] Aravamudham Raman 2007 *Materials Selection and Applications in Mechanical Engineering* vol

40 1st ed (New York: Industrial Press, Inc.)

[5] Gaucher A, Guilhot G and Amsallem C 1976 Surface treatment for iron and steel *Tribol. Int.* **9** 131–7

[6] Mueller J 1969 *United States Patent 3635768: Molten salt carburizing process and replenishing salt for same* 3635768.

[7] Shimer P 1918 *United States patent 1279457: Case-hardening of metals* 1279457.

[8] Foreman R W, Dubal G P and Fox C R 1979 *US Patent 4,153,481 Cyanide-free carburizing process and composition* 4153481.

[9] Foreman R W, Dubai G P and Fox C R 1979 *UK Patent GB2043699 Application Improvement in cyanide-free carburizing process and composition.*

[10] Kerridge D H 1984 *US Patent 4461655 Fused salt bath composition* 4461655.

[11] Buitkus T J 1967 *US Patent 3,331,708 Electrolytic case hardening* 3331708.

[12] Hiyama S 1975 *US Patent 3,876,512 Electrolytic carburizing process using a carbonate electrolyte* 3876512.

[13] Siambun N J 2011 *Electrolysis of Molten Carbonate Salt and Its Application* (Nottingham: University of Nottingham)

[14] Popov K, Djokic S and Grgur B 2002 *Fundamental Aspects of Electrometallurgy* (Boston: Kluwer Academic Publishers)

[15] Oberg E, Jones F D and Ryffel H L H H H 2016 *Machinery's Handbook 30th Edition* (New York: Industrial Press)

[16] Siambun N J, Jewell D A and Chen G Z 2015 Influence of CO_2 Gas in the electro-carburisation process of mild steel *Int. J. Chem. Eng. Appl.* **6**(5) 341–45

[17] Liew W Y H, Ling J L J and Siambun N J 2016 Sliding wear behaviour of steel carburized using Na_2CO_3-NaCl *MATEC Web Conf.* 87

[18] Liew W Y H, Protasius R, Ling J L J, Siambun N J and Mohd-Lair N A 2016 Reciprocating wear behavior of mild steel carburized using Na_2CO_3-NaCl *Tribol. Int.* **95** 406–18

[19] Siambun N J, Liew W Y H, Chen G Z, Jewell D A and Beng Y K 2012 The effect of cooling rate in molten salt electro-carburisation process *Adv. Mater. Res.* **576** 264–7

[20] Bahrin S A, Liew W L Y, Janaun J A, Uvarajan S K A and Siambun N J 2016 The influence ofquenching medium on the hardness and microstructure of electro-carburised low carbon steel in *The 1st Conference on Engineering, Technology & Education 2016* 148–54

[21] Torkamani J R H, Raygan S, and Rassizadehghani J 2014 Comparing microstructure and mechanical properties of AISI D2 steel after bright hardening and oil quenching *Material and Design* **54** 1049–55

[22] Benedetti M, Fontanari V, Höhn B R, Oster P and Tobie T 2002 Influence of shot peening on bending tooth fatigue limit of case hardened gears *Int. J. Fatigue* **24**(11) 1127–36

IOP Conf. Series: Materials Science and Engineering 1217 (2022) 012011 doi:10.1088/1757-899X/1217/1/012011

Fracture energy for orthogonal cutting in unidirectional CFRP at different cutting directions

A N Amir[1], H Ghazali[2,4*], H Wang[3], L Ye[3], N A Fadil[4], W F F W Ali[4], K Ghazali[5] and R Yusoff[5]

[1]Centre of Advanced Materials and Intelligent Manufacturing, Faculty of Engineering, Built Environment & IT, SEGi University Sdn Bhd, 47810 Petaling Jaya, Selangor, Malaysia
[2] Centre for Advanced Composite Materials (CACM), Universiti Teknologi Malaysia, 81310 Johor Bahru, Johor, Malaysia
[3] Centre for Advance Materials Technology (CAMT), School of Aerospace, Mechanical and Mechatronic Engineering, The University of Sydney, NSW 2006, Australia.
[4] School of Mechanical Engineering, Universiti Teknologi Malaysia, 81310 Johor Bahru, Johor, Malaysia
[5] Faculty of Science and Natural Resources, Universiti Malaysia Sabah, 88400 Kota Kinabalu, Sabah, Malaysia

*Corresponding author: habibahghazali@utm.my

Abstract. A unidirectional carbon fibre reinforced polymer (CFRP) laminate is a composite material made up of strong parallel carbon fibres incorporated in a polymer matrix such as epoxy to provide high stiffness and strength in the fibre direction of the laminate. Unfortunately, the interlaminar or intralaminar plane of this material has a low resistance to damages as the fracture toughness of a unidirectional CFRP laminate is related to the energy dissipation during the orthogonal cutting. The aim of this study is on cutting a unidirectional CFRP along the longitudinal or transverse directions, characterizing orthogonal cutting forces and the related fracture energy. Orthogonal cutting is performed using braised carbide tools for a range of cutting depth of 10-100 μm with a rake angle of 30° to quantify the cutting forces and to observe the fracture mechanisms. The fibre orientations have a significant impact on surface bouncing-back. For some fibre orientations, the energy balance model is applicable, deducting the reasonable value of fracture toughness due to high normal force (F_t). Fibre subsurface damage and cutting forces during cutting are found to be strongly influenced by the cutting depth. The input energy of cutting is released in form of new surface energy, fibre breakage, high bending energy, and chip fracture energy.

1. Introduction

Carbon fibre reinforced polymer (CFRP) is one of the composite materials that is durable, low density and is a high corrosion resistant material. CFRP possesses extraordinary properties and has been widely used by the aerospace industry, although it is known that these materials are prone to developing internal damage such as microcracks [1, 2]. This can lead to premature catastrophic failures that can be particularly dangerous for structural stability [3]. The use of CFRP requires a high standard of

Content from this work may be used under the terms of the Creative Commons Attribution 3.0 licence. Any further distribution of this work must maintain attribution to the author(s) and the title of the work, journal citation and DOI.
Published under licence by IOP Publishing Ltd

dimensional accuracy and surface integrity. Compared to isotropic material such as metals, the fracture toughness characterisation of composite materials such as CFRP is still developing due to its anisotropic behaviour.

Most cracking and fracturing cases occur in CFRP during service are Mode-I fracture (opening mode) which involves a tensile stress pulling the crack surfaces apart. Mode-I intralaminar (matrix cracking) fracture toughness is usually measured using double cantilever beam (DCB) specimens according to modified beam theory [4].

The present work aims to determine the Mode-I intralaminar fracture toughness as part of the investigation into the fracture behaviour of CFRP. The energy dissipating mechanisms in orthogonal cutting of UD-CFRP composites is evaluated. Experiments on orthogonal cutting of UD-CFRP surfaces along perpendicular and parallel fibre directions have been conducted. Essential parameters such as specific cutting forces, chips thicknesses are measured to be able to establish fracture toughness from orthogonal cutting. The Mode-I intralaminar cracks grow by creation of a fine intralaminar pre-crack on the surface of the sample. This allows an in-depth investigation on the fibre cutting angles. The fracture toughness in Mode in (G_{IC}) was determined by using combined methods to cater to the relevance of this study.

2. Methodology

The CFRP laminate is pre-made by Composites Technology Research Malaysia (CTRM) consist of 16 layers of UD-CFRP. The laminate was waterjet cut into workpiece with dimension of $50\ mm \times 40\ mm\ \times 2.5\ mm$ with desired fibre orientation of 0° for orthogonal cutting experiments. The specimen with cutting direction and tool tipped is parallel to the fibre direction and the cutting is done layer by layer is called layered specimen. While the specimen with orthogonal cutting in transverse direction and the tool tipped is perpendicular to the fibre direction is called transverse specimen. Figure 1 shows the schematic view of fibre orientation vs. the cutting direction for both layered and transverse specimens. The nominal cutting depth of the test was controlled by CNC machine programming.

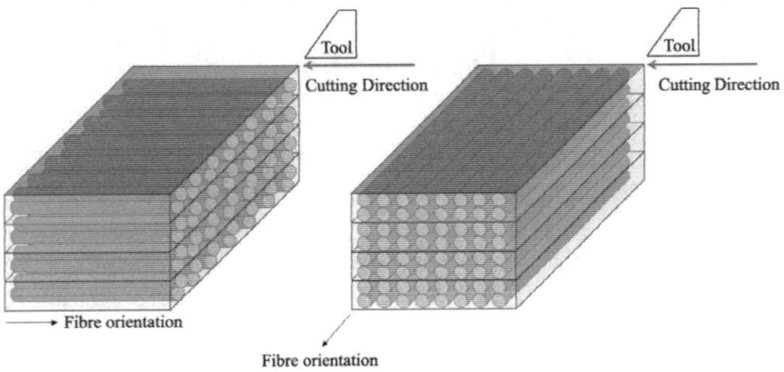

Figure 1. Fibre orientation for orthogonal cutting indicated on the red surface. Left to right: layered and transverse samples.

The orthogonal cutting is operated using Minimi CNC- M 286 Surface Grinder due to its high rigidity (0.12 kN/μm) and precise position control to an accuracy of 0.1 μm. Braised carbide tipped cutting tool with 30° of rake angle and natural sharp cutting edge with a radius less than 7.5 μm is used and mounted on a three-axis piezoelectric force transducer (Kistler 9257B), which measures the cutting forces in horizontal (F_c) and vertical (F_t) direction with 10 mm/sec cutting speed. The cutting depth ranges from 10-100 μm. An accurate measurement was performed with the aid of a chromatic distance sensor (Stil CHR 150), which has a measurement range up to 300 μm with a resolution of 80 nm. The sensor scanned the material cutting surface at different position before and after each cut. The difference between the scans was then calculated for the real depth of cut. Values of cutting force and transverse force refer to the mean values of force in the steady cutting stage. Besides, thermal camera (Micro-Epsilon TIM 160) has been used to monitor the temperature change in the material. Figure 2 shows the experimental setup for the orthogonal cutting.

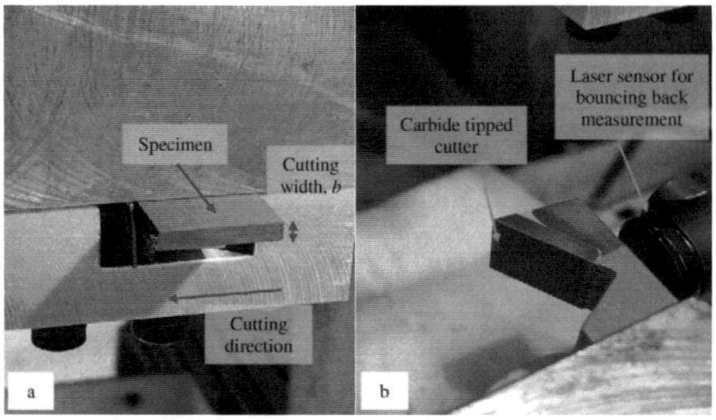

Figure 2. Orthogonal cutting experiment set-up.

3. Results and discussion

3.1 Specific cutting force and thrust force

In this study the cutting forces parallel and perpendicular to the cutting direction are called the *cutting force*, F_C for horizontal force and *transverse force*, F_t for vertical force. The schematic diagram of the machining process and those forces acting on the shear plane is as shown in figure 3(a). During the machining of UD-CFRP parallel to the fibres, the horizontal cutting force (parallel to the fibres and the cutting direction) is determined by the cutting depth and the rake angle, while the vertical cutting force is determined by the relief angle and tool wear. An undeformed chip thickness leads to increases of specific energy due to higher rubbing/friction forces and clearance (relief) of the face of the tool causing increased of the chip thickness h_c [5]. Observation of separation of the top layer between the fibres and matrix as shown in figure 3 is due to the high cutting force pushing the laminate and delamination occurred. This scenario visibly occurs with the transverse fibre orientation.

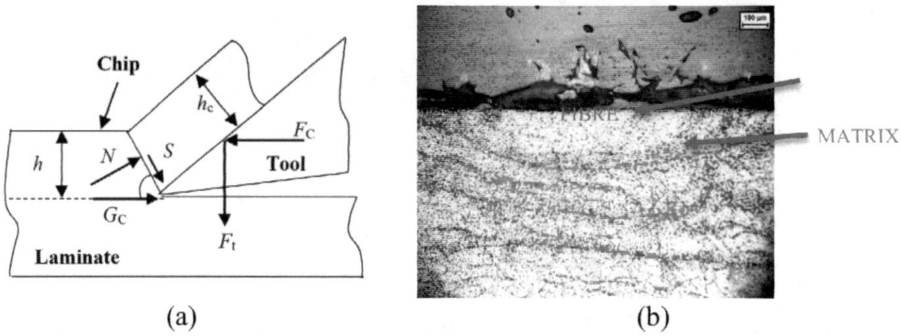

Figure 3. (a) Schematic diagram of the machining process and those forces acting on the shear plane (b) Optical microscope image of the separation of the top layer between the fibres and matrix.

Figure 4 shows that at any fibre orientation, F_t is always higher than F_C. In fact, F_C is consistent and remains low in any orientation. F_t produced by laminates with layer orientation during cutting is unusual as the magnitude of the force is huge with negative value where this could be due to the diameter of the cutter affecting fibre breakage. Fibres are pushed down, triggering the entire structure to move and shift. The internal structure of the laminate experiences softening due to the high cutting force in the perpendicular F_t direction, forcing the fibres to indent. A slightly different cutting mechanism occurs in the transverse direction, where it seems like the fibres are being pushed by the cutters instead of being cut. Therefore, F_t happens to be much smaller compared to the other orientation. A key factor that contributes to the generation of the cutting forces is believed to be the bouncing back phenomenon.

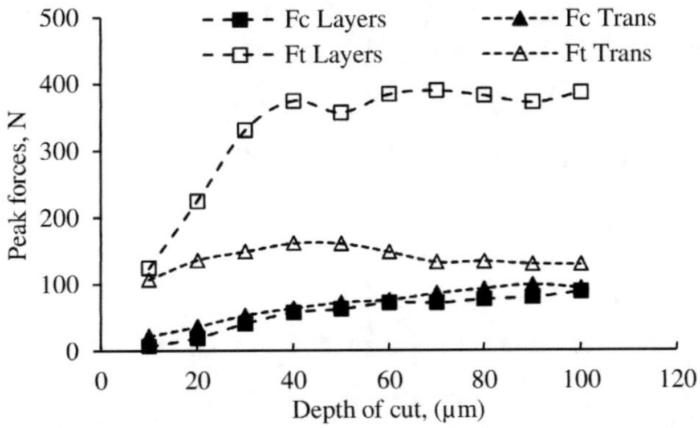

Figure 4. Orthogonal cutting forces F_c and F_t versus depth of cut at different fibre orientations: layer and transverse.

3.2 Influence of bouncing-back phenomenon on orthogonal cutting

The bouncing-back phenomenon is defined as uncut fibres bouncing back after a tool passing through them for a surface removal (cutting) process. Based on orthogonal cutting, it was found that the real depth of cut, and the nominal depth of cut are different. When the tool passed through the sample, the material in the cutting path was pushed down and partially sprang back elastically. It was also found that the magnitude of bouncing back was related to the radius of the cutting tool. The diameter of the fibre used was 6.5 ± 0.4 μm and the sharpness of the tool tip was 7.1 μm. An assemblage of fibres that have been cut might experience unfinished cutting due to fibres bending, leading to the bouncing-back phenomenon. Figure 5 (a) shows a critical analysis between nominal cutting depth, real cutting depth and bouncing-back height. When the cutting tool passed through, fibres recovered elastically as many fibres were pushed and caused to bend (but did not break at the cutting point) (figure 5 (b)), causing the magnitude of bouncing-back to become larger. These results indicate that the nominal cutting depth will always be different than the real cutting depth. The real cutting depth is defined as the difference in the surface distances measured by the laser sensor before and after cutting.

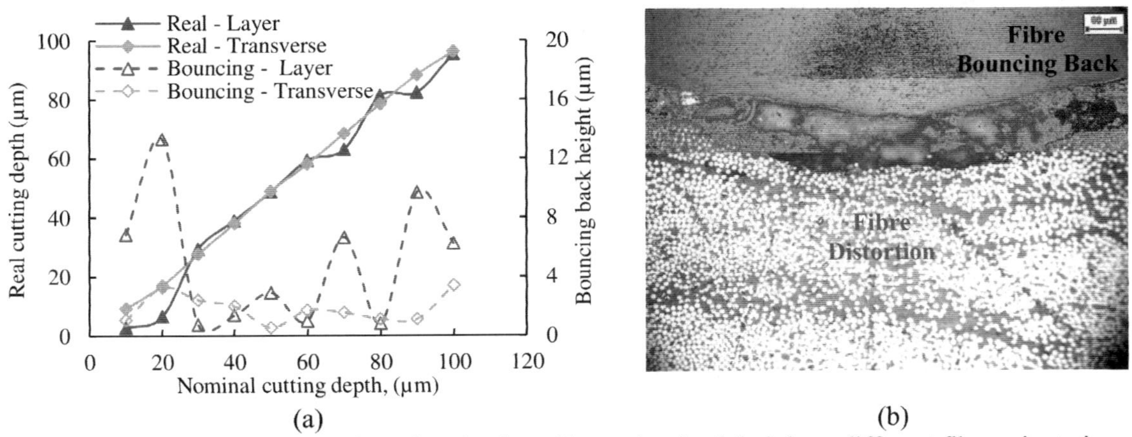

Figure 5. a) Nominal versus real cutting depth and bouncing back height at different fibre orientations b) microstructure of fibre distortion and fibre bouncing back phenomena in cut laminate.

3.3 Determination of G_C using energy balance model for cutting

It is believed that the crack initiation itself is not the cause of fractures in composites. Fractures occur when a defect larger than a threshold size is formed [6]. Fracture toughness for cutting was determined using combined methods to cater to the relevance of this study. According to Patel *et al.* [7], when cutting reaches a steady state, the energy balance is the total external work done by the cutting tool, which equals the sum of the plastic work on the shear plane, the frictional loss at the tool-chip interface, and the work spent in fracturing the material [8]. For this study, a modified method was used to determine fracture toughness.

Modifying William's energy balance model by using coefficient of friction, μ:

$$\tan\varphi = \frac{\cos\alpha}{\frac{h_c}{h} - \sin\alpha} \tag{1}$$

φ is the angle of the shear plane that is related to tool rake angle α and chip thickness h_c, where h is the cutting depth,

$$\frac{F_c}{b} - \frac{F_t}{b} tan\,\phi = \frac{\sigma_y h}{2}\left(tan\,\phi + \frac{1}{tan\,\phi}\right) + G_c \tag{2}$$

Thus, a simplified equation from Merchant's force minimization criterion and, G_c and σ_y can be determined from linear plot of $(F_c/b - F_t/b\tan\varphi)$ versus $(h/2)(\tan\varphi + 1/\tan\varphi)$ data from equation (2). The gradient gives σ_y and the positive intercept on the Y-axis is G_c and b is the thickness of the specimen.

By using known σ_y, which is determined by the gradient plot of equation (2),

$$\cot\phi = \sqrt{1 + \frac{2}{\sigma_y h}\frac{F_t}{b}} \tag{3}$$

and,

$$Z = \frac{\mu - \tan\alpha}{1 + \mu\tan\alpha} \tag{4}$$

where Z is the friction factor and coefficient of friction, $\mu = 0.41$. S is the shear force and N is the normal force on the tool-chip interface, during cutting and machining, heat is often developed along the tool-chip interface even though the value is negligible, yet it leads to chip adhering to the tool and an adhesion toughness, G_a is required. By including the adhesion toughness and coefficient of friction (COF), tool-chip interface is as below:

$$\frac{S}{b} - G_a = \frac{N}{b} \tag{5}$$

$$\frac{F_c}{b} = \sigma_y h.\cot\phi + G_c \tag{6}$$

Thus, G_C and σ_y can be determined from linear plot of (F_c/b) versus $(h \cdot \cot\phi)$. The gradient gives σ_y and the positive intercept on the Y-axis is G_C.

Chip thickness h_c will always larger than cutting depth h due to the shearing effect during cutting. As composites contain both fibre and matrix materials, polymer characteristics are relevant during cutting. Bending terms in the energy balance are controlled by this characteristic [8, 9]. Figure 6 shows a linear plot from equation (2) that gives a consistent pattern of results at any depth of cut. Fibre in the transverse direction produces a smaller value of fracture toughness compared to layered orientation. Table 1 shows the value of fracture toughness and yield stress obtained from the plot. This modified method is generated based on the original William's energy balance model which resulting reliable results from figure 6. In this case, a modified method gives a better analysis of orthogonal cutting at different UD fibre orientations as it includes friction energy, and this process will involve friction at the tool-chip interface.

Table 1. G_{Ic} and σ_y values obtained from modified energy balance model.

Fibre orientation	Method: Coefficient of friction (kJ/m^2)	
	Fracture toughness, G_{Ic} [kJ/m^2]	Yield stress, σ_y [MPa]
Layered	5.8	66
Transverse	5.2	103

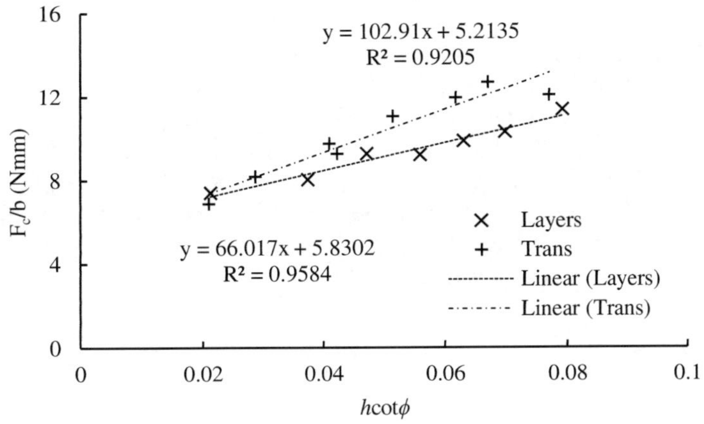

Figure 6. (F_c/b) versus ($h \cdot \cot \phi$) plot.

4. Conclusion

The aim of this study is to determine the Mode-I intralaminar fracture toughness as part of the investigation into the fracture behaviour of CFRP. The energy dissipating mechanisms in orthogonal cutting of UD-CFRP composites has been evaluated. Experiments on orthogonal cutting of UD-CFRP surfaces along perpendicular and parallel fibre directions have been conducted. Conclusions drawn from this experimental study are as follow.

1. The results of the present study show that the average critical energy release rates, G_{IC}, indicating that the method based on orthogonal cutting can produce a reliable as positive fracture toughness value for intra-laminar fractures.
2. Bouncing back phenomenon is always influenced by the orientation of the fibres. Due to this phenomenon nominal cutting depth will always be different than real cutting depth.
3. The results demonstrated that intra-laminar fracture toughness depends on fibre orientation. Fibre in the transverse direction produces a smaller value of fracture toughness compared to layered orientation.

Acknowledgement

The author(s) disclosed receipt of the following financial support for the research, authorship, and/or publication of this article: This work was supported by Universiti Teknologi Malaysia, UTMER grant, cost center no. Q.J130000.3851.19J39

References

[1] Borstnar G, Mavrogordato M N, Helfen L, Sinclair I and Spearing S M 2015 Interlaminar fracture micro-mechanisms in toughened carbon fibre reinforced plastics investigated via synchrotron radiation computed tomography and laminography *Composites Part A: Applied Science and Manufacturing* **71** 176-83

IOP Conf. Series: Materials Science and Engineering 1217 (2022) 012011 doi:10.1088/1757-899X/1217/1/012011

[2] Su C, Wang X, Ding L and Yu P 2021 Enhancement of mechanical behavior of resin matrices and fiber reinforced polymer composites by incorporation of multi-wall carbon nanotubes *Polymer Testing* **96** 107077

[3] Ghazali H, Ye L and Zhang M Q 2017 Mode II interlaminar fracture toughness of CF/EP composite containing microencapsulated healing resins *Composites Science and Technology* **142** 275-85

[4] ASTM D 5528-01 *Standard Test Method for Mode I Interlaminar Fracture Toughness of Unidirectional Fiber-Reinforced Polymer Matrix Composites* (West Conshohocken: ASTM International)

[5] Armarego E and Brown R 1961 On the size effect in metal cutting *The International Journal of Production Research* **1**(3) 75-99

[6] Hwang W, Park Y and Hwang W 2005 Strength of fiber reinforced metal laminates with a circular hole *Metals and Materials International* **11**(3) 197-204

[7] Patel Y, Blackman B and Williams J 2009 Determining fracture toughness from cutting tests on polymers *Engineering Fracture Mechanics* **76**(18) 2711-30

[8] Wang H, Chang L, Ye L and Williams J 2015 On the toughness measurement for ductile polymers by orthogonal cutting Engineering *Fracture Mechanics* **149** 276-86

[9] Wang H, Amir FA, Chang L and Ye L. On the Determination of Interlaminar Fracture Toughness of Fibre Reinforced Polymer Laminates Using Orthogonal Cutting *21st International Conference on Composite Materials (ICCM-21)* 20-25th August 2017 Xi'an China

IOP Conf. Series: Materials Science and Engineering 1217 (2022) 012012 doi:10.1088/1757-899X/1217/1/012012

Investigation of specific energy absorption of natural fiber honeycomb sandwich structure composite as building construction material application

Nahiyan Al-Azad*, Mohd. Kamal Mohd Shah

Advanced Composite and Material Research Group (Environmental and Health Technology), Faculty of Engineering, Universiti Malaysia Sabah

*Corresponding authors: nahiyanalazad@gmail.com

Abstract. The usage of incorporating natural fibre in composite material has seen some potential to be used as a future building construction material due to its recyclability, lightweight and high-reliability feature. However, the issue of implementing natural fibre as building construction material in composites material concerns the structural integrity of the material. As the characteristics of the natural fibre honeycomb composite have been discovered more in terms of properties which ranges from its physical and chemical structural composition to the quasi-static impact collapse of the material, the absorption energy of the material in different cell geometry is unknown. Therefore, the purpose of this study involves the testing of the natural fibre honeycomb (NFH) composite made from cement fibre (face sheet) and corn starch (core) with regards to its crushing behaviour when subjected to flatwise compression load according to ASTM D-3410 standard to analyse the performance of energy absorption of NFH composite with different thicknesses of the hexagonal core and cell wall thickness to determine the Specific Energy Absorption of the material. The result obtained shows that the increasing thicknesses of the core and cell wall improves the ability of the composite to absorb more energy and the specific energy absorption is higher when both factors are increased.

1. Introduction

Natural fibre composites (NFC) are low-cost materials offering a low density, high specific properties, biodegradable characteristics, and recyclability. Lightweight composite structures have been seen in many applications such as aircraft and satellites due to the great weight reduction output. Composite sandwich structures are used to offer a high strength to weight ratio as well as high specific bending stiffness and strength characteristics under distributed loads.

The interest in NFC materials is drastically increasing both in terms of their industrial applications and fundamental research. Many researchers found them as a flexible material to be incorporated as composites as they are renewable, low cost, corrosion-resistant, easy to manufacture, completely or partially recyclable, and biodegradable [1 – 4]. These fibres originate from plants, such as flax, cotton, hemp, jute, kenaf, palm, bamboo, as well as wood have been used since the sixties and are more often applied as reinforcement of composites [5, 6]. Their availability, renewability, low density, and price as well as mechanical properties make them one of the best ecological alternatives to glass, carbon, and man-made fibres commonly used for the manufacturing of composites.

Content from this work may be used under the terms of the Creative Commons Attribution 3.0 licence. Any further distribution of this work must maintain attribution to the author(s) and the title of the work, journal citation and DOI.
Published under licence by IOP Publishing Ltd

While most researchers highlighted the usage of NFC in the automobile and aeronautics field, reference [7] focused on the usage of the NFC as future construction material. The research focused on the compression test of NFC which incorporates flax, jute, and hemp while considering the fibre and resin design, and structural geometry of NFC. The initial findings of structural properties and design methods show positive results for residential and light commercial building applications; however, the author highlighted that further research is necessary [7].

The characteristics of the NFH composite have been more discovered in terms of properties which ranges from its physical and chemical structural composition to the quasi-static impact collapse of the material [3 – 7]. The specific absorption energy (SEA) of NFH in different certain cell geometry was found to be unknown and is yet to be investigated. The study on SEA is a crucial element in the design of parts that require weight reduction, such as land or aerial vehicles. However, this study has seen to be applied more in the automobile industry [8 – 10]. The construction industry currently uses heavy materials to construct buildings, and this was seen to cause the construction time to be longer [11]. Therefore, a need for a lightweight material in construction industry was highlighted in several studies previously [9, 10, 12].

This research identifies the crushing behaviour of natural fibre honeycomb composite which is made from wood pulp fibre (face sheet) and corn starch (core) with regards to its core thickness when the material is subjected to flatwise compression load. The result obtained from the compression test will then be used to further analyse the performance of energy absorption of Natural Fibre Honeycomb (NFH) composite with different thicknesses of the hexagonal core and cell wall thickness.

2. Methodology

The sample was prefabricated from a factory located in Jiangsu, China. The core material was designed in hexagonal shape and manufactured by using the corrugation method. A honeycomb-patterned core with density 1200 kg m^{-3} and made off cement-fibre with wood pulp fibre (face sheet) and corn starch mixed with cardboard (core) (figure 1). Figure 2 and figure 3 shows the image of the sample after fabrication. There are two (2) different core cell wall thicknesses 0.145 mm and 0.29 mm were selected. The sample was cut into different dimensions as shown in table 1. To make sure the reliability of test data, a larger size within the range that equipment would allow was chosen.

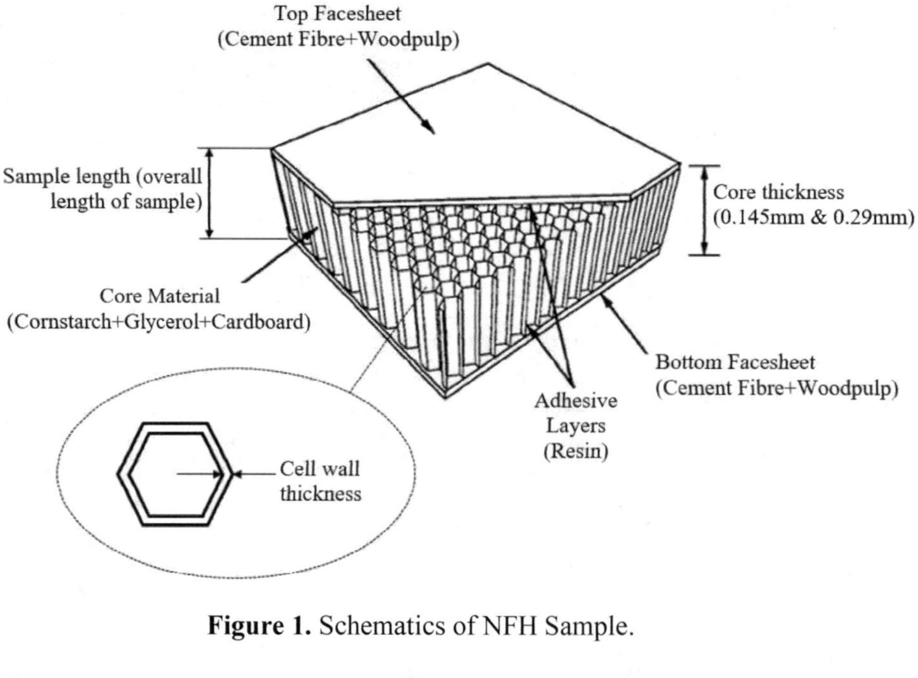

Figure 1. Schematics of NFH Sample.

Figure 2. Overall view of NFH sample.

Figure 3. View of NFH sample's core.

Table 1. Sample dimensions.

Type	Sample Code	Cell Wall Thickness (mm)	Core Thickness (mm)	Sample Length (mm)
I	O120	0.145	90	120
	O80		60	80
	O60		45	60
II	M120	0.29	90	120
	M80		60	80
	M60		45	60

A flat-wise compression test was performed according to the ASTM D-3410 standard by using Shimadzu Universal Tester Machine (figure 4) to determine the in-plane compressive properties of polymer matrix composite materials reinforced by high modulus on natural fibre honeycomb (NFH) composite with a cell wall thickness of 0.145 mm and 0.29 mm. The compression loading is applied to the sample with a crosshead speed of 5 mm min^{-1} until failure of the core occurs. Each sample was repeated three (3) times to obtain an average reading to minimize the error that may occur during experiment testing. The reading obtained from the test represents the energy absorption characteristics of the sample when exerted to axial loading. Based on the graph that will be obtained after the compression test, the area under the curve represents the total absorbed energy absorbed (equation 1) during the plastic deformation of the NFH composite.

$$`E_{absorbed} = \int P \ \delta l \tag{1}$$

Where P is the force load and δl is the compressive displacement.

The amount of energy absorbed by each unit mass of sample or also known as specific energy absorption (SEA) is obtained by the following equation 2 where the result of equation (1) is divided by the mass of the sample, which is denoted by, M.

$$SEA = \frac{E_{absorbed}}{M} \tag{2}$$

Figure 4. Shimadzu Universal Tester Machine.

3. Result and discussion

3.1 Type I sample

The axial loading-displacement curve is shown in figure 5 below refers to the result obtained for the type I sample. The standard deviation obtained from the type I sample is between 2.16-2.71 which indicates that the values are well spread out. There are 3 typical stages of failure that usually occur in the compressive test. Stage I failure is known as the elastic stage. In this stage, an in-plane compression of the core was the contribution to load carrying, and no damage would happen before the peak compressive force. Stage II failure is known as the collapse stage. Stage III is known as the densification stage where after the complete collapse of the sample, the load begins to rise again. In this experiment, both samples exhibited the 3 stages of failure. However, figure 5 shows that overall, the material has undergone ductile failure.

Based on figure 5, for the sample code O120, there is a total of three peaks of axial loading observed throughout the axial loading test. They have occurred during 5 mm compression displacement where the Stage I failure occurs, 55 mm compression displacement where Stage II failure ends, and 65 mm compression displacement where Stage III failure was observed. When the load is at the peak, the sample goes through buckling which constructs a stable structure to provide more load. According to reference [13], the increase of loading value in the graph of load against displacement indicates the sample undergoes an elastic deformation process where folding occurs on the sample during the elastic deformation stage. Thus, this indicates that the sample has the potential to absorb more energy. The maximum loading where sample O120 was stable occurred during 4mm displacement at 10.36 kN load.

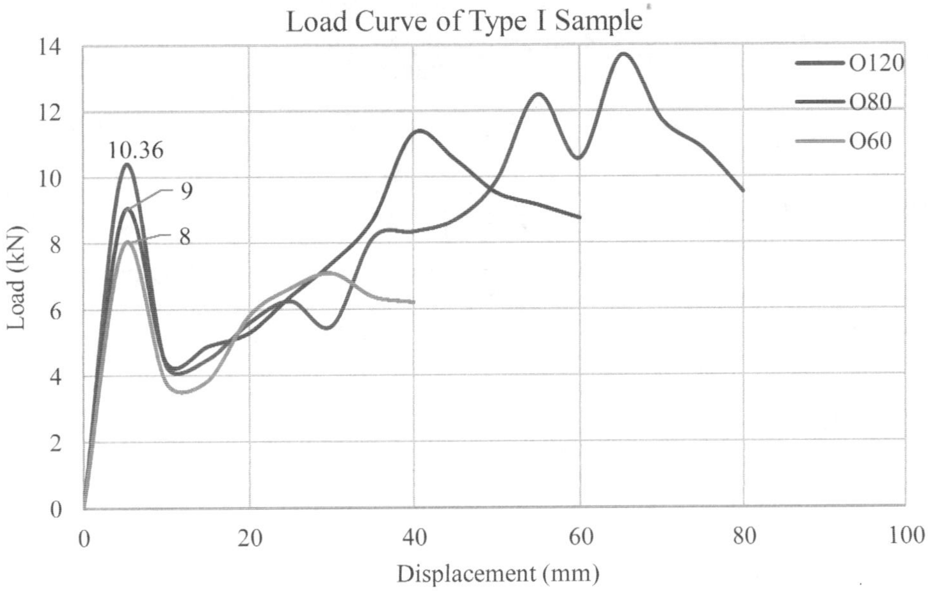

Figure 5. The load-displacement curve for type I sample.

For the sample code O80, it is observed in figure 5, there are two peaks of axial loading value throughout the axial loading test which occurs at 9 mm and 41 mm of displacement. Elastic deformation of sample O80 was observed during compression displacement of 4 mm. The folding of the sample started to occur when the displacement reaches 4 mm at the upper part of the sample while at the lower part of sample O80 folds at 9 mm displacement before it reaches to densification stage where it's completely folded. After the sample achieved the maximum load stage, there are no bending at the cell walls observed, the system collapsed at the lower part and a large drop in the load level occurred. It is believed that the cause of the drop is due to the bending and local buckling of the cell walls. The load continued to increase with a small slope after the initial drop. The increase in load capacity is caused by the densification of the folded cell walls. The maximum loading where the sample was observed to be stable structurally was at 9 kN at 4mm compression.

For the case of sample O60, the 0 to 4mm displacement region in figure 5 indicates the elastic stage, where the sample absorbs the most energy during the deformation process. Therefore, this stage determines the honeycomb's mechanical properties where maximum load it could absorb before failure of the core was at 8 kN. When it comes to point 25 mm to 30 mm displacement, the stress on the sample will increase rapidly due to densification at the same time, the honeycomb structure loses its buffering property. Therefore, it can be concluded that the honeycomb structure's stress rises rapidly during the elastic stage. This phenomenon depicts the approximate ideal linear elastic relationship.

3.2 Type II sample
Figure 6 below shows the load against displacement of sample type II. Type II sample had a standard deviation of 2.15-3.97 which indicates that the values obtained are well spread out. A similar pattern can be observed for the Type II sample where the failure occurs in 3 stages which are Stage I which is elastic, Stage II which is collapsing stage, and Stage 3 which is the densification stage. Ductile failure was observed at the end of the experiment. For sample code M120, it can be observed that the first local buckling occurs on the 5 mm displacement at 18.13 kN. The increase of loading value in the curve indicates the sample undergoes a plastic deformation process where the sample's folding occurs during the plastic deformation stage therefore, it indicates there is a potential for the material to absorb more energy.

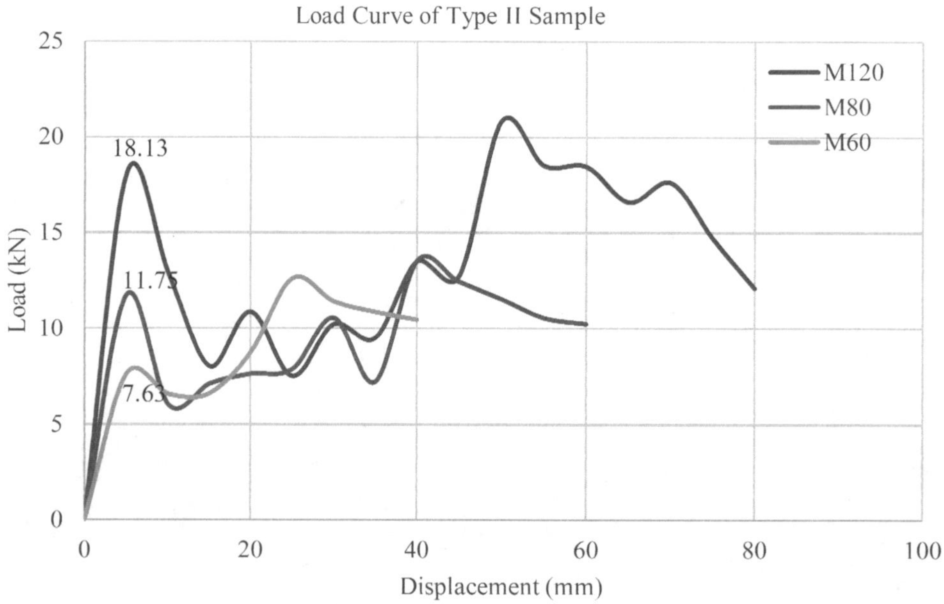

Figure 6. The load-displacement curve for the type II sample.

The sample M80 in figure 6 shows there is a total of three peak axial loading values throughout the test. They have occurred during 5 mm, 30 mm, and 40 mm compression displacement. The first local buckling occurred on 30 mm when the sample can no longer withstand the force exerted on the sample. In this stage, some amount of energy is dissipated from the sample thus it explains the drastic drop in the loading. The second folding was seen to occur during 40 mm compression displacement. The maximum loading where sample M80 was observed to be stable was at 11.57 kN during 5mm displacement.

Sample M60 on the other hand exhibited the lowest among both Type I and Type II sample. The maximum loading that M60 was capable to withstand without failure was at 7.63 kN at 5 mm displacement. However, it was capable to absorb more energy until it reaches densification stage at 2 mm displacement.

3.3 Specific absorption energy (SEA)

Based on the graph obtained from figure 5 and figure 6, the SEA of all the samples was obtained through equation (2) with the results from equation (1). Table 2 below shows the tabulated result obtained. Based on table 2, a graph of comparison for SEA was obtained as depicted in figure 7 below.

Table 2. Tabulated result for SEA.

Sample Code	Average SEA (J)	Mass (kg)	SEA (Jkg^{-1})
	Type I		
O120	686.52	0.69	994.96
O80	452.42	0.46	983.52
O60	226.18	0.35	646.23
	Type II		
M120	1115.65	0.69	1616.88
M80	530.79	0.46	1153.89
M60	367.77	0.35	1050.77

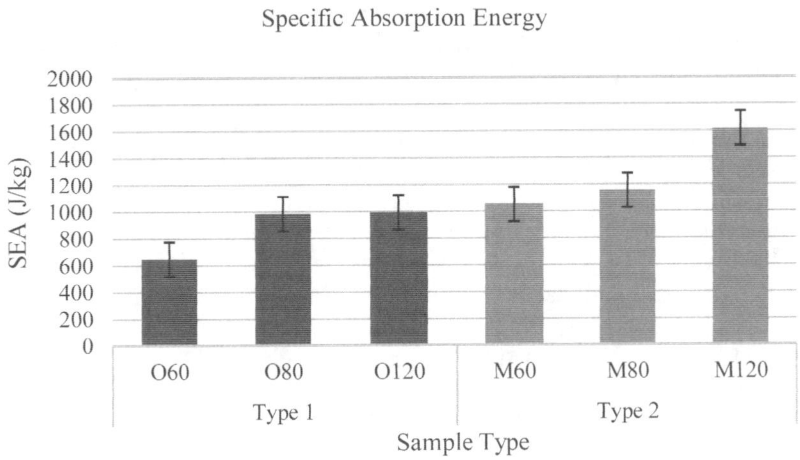

Figure 7. SEA comparison of type I and type II sample.

From figure 7 above, an observation was seen for both types of samples (Type I and Type II) on the difference of SEA value between the sample of the same material but different core thickness. It was observed that the SEA increases as the sample's core thickness increases. This behaviour is in line with the results from reference [14]. The SEA increases with core thickness of the sample because the mass of the cell wall increases as the cell wall gets thicker [14 – 16]. However, according to reference [14], it is believed that wall thickness is not a major cause for change in the overall deformation modes. The type II sample depicted that the SEA increases moderately as the sample length increases. While, for sample Type I, there is SEA value consistency spotted when the sample length increases from 80mm to 120 mm. This is an expected trend according to a study performed in reference [14]. Apart from that, it is noted that the variance amount of SEA value between two samples of the same length, but different cell wall thickness increases as the sample length increase. A huge difference in the amount of SEA value occurred between sample O120 and M120 where the SEA value for O120 is found to be

994.96J kg^{-1} while the SEA value for M120 is found to be 1616.88J kg^{-1}. Thus, it can be deduced that sample type II exhibits a high capacity to absorb more energy compared to sample Type I.

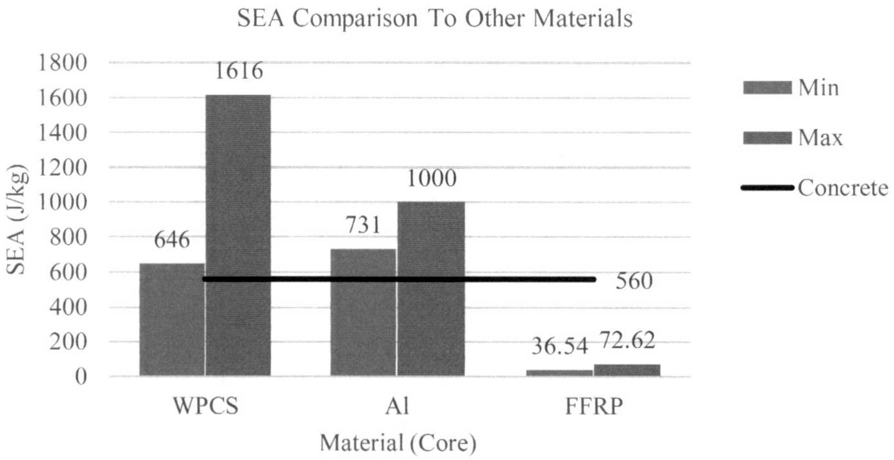

Figure 8. SEA comparison of the core to other materials.

Figure 8 above shows the comparison of the SEA of the core as compared to other research by reference [15] and [17] that uses a similar motive. The conventional material, concrete, which is currently being used has been depicted in figure 8 as the benchmark for the SEA value [18]. The results from these experiments were found to be significantly greater than the study performed in reference [15]. The author performed a similar study to analyse the SEA of flax fibre reinforced polypropylene (FFRP) which is commonly used nowadays. It was found in reference [15] that materials with the greater structure sizes are exhibits a capability to absorb a higher amount of crushing energy. The SEA value by FFRP was recorded between 36.54-72.62 Jkg^{-1}. Another study with a similar motive was studied in reference [17] where it was found that for aluminium (Al) sandwich structure core, the SEA value was found to be 1000 Jkg^{-1} on average. This has been deduced that a combination of wood pulp fibre and corn starch (WPCS) which has been used in this research is far better than reference [15] and [17]. As concrete is currently used as construction material for buildings, based on figure 7, it can be deduced that WPCS as core material in sandwiched structure composite is capable to be used for building material from a SEA point of view.

4. Conclusion

The energy absorption performance of each sample was determined by subjecting it under compression loading. The mechanical behaviour was identified under the influence of the geometrical properties of the honeycomb core. Application of flatwise compression tests to the honeycomb core material showed that energy absorption increases with the core thickness as the honeycomb cell wall thickness increase. During the flatwise compression test, the honeycomb core cell wall buckled locally and densified as the compression displacement increases. Based on the experimental findings, the sample that has the highest value of SEA was sample M120 which has 90 mm core thickness and 0.29 mm cell wall thickness. Therefore, it is concluded that samples with higher core thickness and cell wall thickness increase the value of SEA thus the material has a higher capacity to absorb more energy thus the material does have the potential to be applied in the construction industry as a future building material. By comparing to other materials, WCPS's SEA value confirms that this material can be used as a building material from a SEA point of view provided the core and cell wall height is thick. As this research is still on-going research, further experiments could be carried out to confirm the usage of WCPS in a sandwiched structure composite as a building material such as flexural test, fire structure test, tensile test, and water absorption test.

References

[1] Sailesh A, Arunkumar R and Saravanan S 2018 Mechanical Properties and Wear Properties of Kenaf–Aloe Vera–Jute Fiber Reinforced Natural Fiber Composites *Materials Today: Proceedings* **5**(2) 7184-90

[2] Alzebdeh K, Nassar M and Arunachalam R 2019 Effect of fabrication parameters on strength of natural fiber polypropylene composites: Statistical assessment *Measurement* **146** 195-207

[3] Elanchezhian C, Ramnath B, Ramakrishnan G, Rajendrakumar M, Naveenkumar V and Saravanakumar M 2018 Review on mechanical properties of natural fiber composites *Materials Today: Proceedings* **5**(1) 1785-90

[4] Navaneethakrishnan G, Karthikeyan T, Saravanan S, Selvam V, Parkunam N, Sathishkumar G and Jayakrishnan S 2020 Structural analysis of natural fiber reinforced polymer matrix composite *Materials Today: Proceedings* **21** 7-9

[5] Cavalcanti D, Banea M, Neto J, Lima R, da Silva L and Carbas R 2019 Mechanical characterization of intralaminar natural fibre-reinforced hybrid composites *Composites Part B: Eng.* **175** 107149

[6] Immonen K, Anttila U and Wikström L 2018 Coupling of PLA and bleached softwood kraft pulp (BSKP) for enhanced properties of biocomposites *J. Thermoplastic Composite Mat.* **32**(3) 328-41

[7] Bambach M 2020 Direct Comparison of the Structural Compression Characteristics of Natural and Synthetic Fiber-Epoxy Composites: Flax, Jute, Hemp, Glass and Carbon Fibers *Fibers* **8**(10) 62

[8] Alsubari S, Zuhri M Y, Sapuan S M, Ishak M R, Ilyas R A and Asyraf M R 2021 Potential of natural fiber reinforced polymer composites in sandwich structures: A review on its mechanical properties *Polymers* **13**(3) 423

[9] Amir A L, Ishak M R, Yidris N, Zuhri M Y and Asyraf M R 2021 Potential of Honeycomb-Filled Composite Structure in Composite Cross-Arm Component: A Review on Recent Progress and Its Mechanical Properties *Polymers* **13**(8) 1341

[10] Al-Azad N, Mojutan E C and Shah M K 2021 A Mini Review on Natural Fiber Honeycomb (NFH) Sandwiched Structure Composite: Flexural Perfomance Perspective *J. Mat. Sci. and Chem. Eng.* **9**(5) 1-0

[11] Al-Azad N, Enegbuma W I, Bamgbade J A and Ohueri C C 2019 Influencing factors in extension of time claims *Australasian Universities Building Education Association Conference* **43** 70-85

[12] Jarasjarungkiat A and Chuaisrinual S 2019 Development of Effective Lightweight Material for Construction Building *IOP Conference Series: Mat. Sci. and Eng.* **639** 012034

[13] Darvizeh A, Darvizeh M, Ansari R and Meshkinzar A 2013 Effect of low density, low strength polyurethane foam on the energy absorption characteristics of circumferentially grooved thick-walled circular tubes *Thin-Walled Structures* **71** 81-90

[14] Zhu G, Sun G, Yu H, Li S and Li Q 2018 Energy absorption of metal, composite and metal/composite hybrid structures under oblique crushing loading. *Int. J. Mech. Sci.* **135** 458-83

[15] Ibrahim M and Yusoff M 2018 Optimization of Interlocking Structures Made of Flax Fibre Composites to Improve Its Energy Absorption Capability *J. Adv. Res. in App. Sci. and Eng. Tech.* **10**(1) 1-17

[16] Yang X, Ma J, Shi Y, Sun Y and Yang J 2017 Crashworthiness investigation of the bio-inspired bi-directionally corrugated core sandwich panel under quasi-static crushing load *Mat. & Design* **135** 275-90

[17] Roslan S, Hassan M, Rasid Z, Zaki S, Daud Y, Aziz S, Sarip S, Ismail Z 2015 Mechanical properties of bamboo reinforced epoxy sandwich structure composites *Int. J. Auto. and Mech. Eng.* **12** 2882-92
[18] Wang X, Liu L, Zhou H, Song T, Qiao Q and Zhang H 2021 Improving the compressive performance of foam concrete with ceramsite: Experimental and meso-scale numerical investigation *Mat. & Design* 109938

Methane hydrogen laminar burning velocity blending laws in horizontal open-ended flame tube rig

N A Amaludin[1*], M Morrow[2], R Woolley[2], A E Amaludin[3*]

[1]Centre of Research in Energy and Advanced Materials, Faculty of Engineering, Universiti Malaysia Sabah, Kota Kinabalu, 88400, Sabah, Malaysia.
[2]Mechanical Engineering Department, Engineering Heartspace, Sir Frederick Mappin Building, Mappin Street, Sheffield S1 3JD, United Kingdom.
[3]Faculty of Engineering, Universiti Malaysia Sabah, Kota Kinabalu, 88400, Sabah, Malaysia.

[*]Corresponding author: nazrein.amaludin@ums.edu.my, adriana@ums.edu.my

Abstract. Different fuel properties and chemical kinetics of two different fuels would make it challenging to predict the combustion parameters of a binary fuel. Understanding the effect of blending methane and hydrogen gas is the main focus of this paper. Utilizing a horizontal tube combustion rig, methane-hydrogen fuel blends were created using blending laws from past literature, over a range of equivalence ratios from 0.6 – 1.2 were studied, while keeping one combustion parameter constant, the theoretical laminar burning velocity. The selected theoretical laminar burning velocity for all the mixtures tested were kept constant at 0.6 ms^{-1}. Different factors affected the flame propagation across the tube, including acoustic pressure oscillations, heat loss from the rig, and obvious difference in hydrogen percentage in the fuel blends. The average experimental laminar burning velocity of all the flames was 0.368 ms^{-1}, compared to the expected value of 0.6 ms^{-1}. In an attempt to keep the theoretical laminar burning velocity constant for different mixtures, it was discovered that this did not promise the same flame propagation behaviour for the tested mixtures. Further experimentation and analysis are required in order to better understand the underlying interaction of the fuel blends.

1. Introduction

Carbon emissions from fossil fuels have been increasing at an alarming rate, which initiated the interest in using alternative fuels to reduce the global carbon footprint. An extensive research and development would be required to design machines that operate on alternative fuels alone due to the obvious difference between fossil fuels and alternative fuels [1]. Apart from that, migration to alternative fuels would leave the current machineries to be left unused, causing further loss to existing fossil fuel users. In order to make the migration to alternative fuels less intrusive, mixing fossil fuels and alternative fuels seems to be the best option, but it will not be an easy task. The different fuel properties and chemical kinetics of the fuels would make it challenging to predict the combustion parameters of the binary fuel [2]. Methane is a preferred hydrocarbon to mix with alternative fuels since it is the cleanest burning fossil fuel, due to its simple structure [3]. On the other hand, hydrogen is an alternative fuel which has one of the highest potential, thanks to its high burning rate and carbon free structure [4].

1.1 Fuel composition

The factors which determine the fuel composition of a binary fuel are the equivalence ratio, ϕ, mole fraction of the first fuel, $\bar{x}_{fi}$, and the mole fraction of the second fuel, $\bar{x}_{fj}$. It is important to note that different fuels require different amount of air in order to oxidize completely, thus the need to include the air required by both fuels in the following manner:

$$\bar{x}_{fi} = \frac{\left[{}^{\phi}/_{(\phi + a_{si})}\right]\bar{x}_i}{\left[{}^{\phi}/_{(\phi + a_{si})}\right]\bar{x}_i + \left[{}^{\phi}/_{(\phi + a_{sj})}\right]\bar{x}_j} \tag{1}$$

where $\bar{x}_f$ = mole fraction of fuel within the binary fuel mixture.
ϕ = equivalence ratio.
a_s = stoichiometric air mole/fuel mole.
$\bar{x}$ = mole fraction fuel within the air-fuel mixture, $\bar{x}_i + \bar{x}_j = 1$
subscript *i* = fuel *i*
subscript *j* = fuel *j*

The normal approach in utilizing equation 1 is by setting value for $\bar{x}_{fi}$ and ϕ to obtain an equation with two unknowns, $\bar{x}_i$ and $\bar{x}_j$. The equation will then be solved using equation $\bar{x}_i + \bar{x}_j = 1$ to obtain the value of $\bar{x}_i$ and $\bar{x}_j$. Fractional mass weightings are calculated by multiplying the mole fractions with the molecular weight of the respective fuels, followed by normalizing the mass weightings to 1. The mole and mass fraction of fuels within the air-fuel mixture will then be used in laminar burning velocity blending laws.

1.2 Laminar burning velocity blending laws

Six different laminar burning velocity blending (u_l) laws were commonly used over the years, which were based on fractional mass weighting of u_l of constituent mixtures law [5], fractional mole weighting of u_l of constituent mixtures [6], modified Le Châtelier law [7], laminar burning velocity plotted against molar heat of reaction ($\bar{Q}$) per mole of constituent mixture [8], fractional mass weighting of constituent mixture $u_l\rho_u$ with the expression of mass rate of burning [9], and finally fractional mole weighting of the activation temperature, $\tilde{T}_a$, used to calculate the $\tilde{T}_a$ of the blend. Details of the blending laws were tabulated in table 1, which shows that the fifth law has the smallest ratio of predicted laminar burning velocity to the actual laminar burning velocity, at 1.01, with a standard deviation of 0.03.

Table 1. Different laminar burning velocity blending laws for binary fuels.

No.	*Laminar Burning Velocity Blending Law*	Ref.	u_{lp}/u_l	σ
1	Fractional mass weighting of u_l of constituent mixtures.	[5]	1.09	0.06
2	Fractional mole weighting of u_l of constituent mixtures.	[6]	1.2	0.09
3	Modified Le Châtelier law. Fractional mole weighting of constituent fuels.	[7]	0.92	0.04
4	u_l plotted against $\bar{Q}$ per mole of constituent mixture.	[8]	1.18	0.08
5	Fractional mass weighting of constituent mixture $u_l\rho_u$ with $(QC_p/k_u)^{0.5}$.	[9]	1.01	0.03
6	Fractional mole weighting of $\tilde{T}_a$ to obtain blend $\tilde{T}_a$.	[10]	0.89	0.04

The current work utilized the fifth law proposed by Bradley *et al.* [9], since it has the best prediction to experimental velocity ratio of 1.01, and standard deviation of 0.03. This law was based on the work of Spalding [11, 12], where the mass of rate of burning was rearranged in the following form:

$$(Qc_p/k_u)^{0.5}u_l\rho_u = \left[\frac{\int_0^1 R(c)dc}{\lambda}\right]^{0.5} \tag{2}$$

$$Q = c_p(T_b - T_u) \tag{3}$$

where
Q = mass heat of reaction (J/kg)
c_p = mean specific heat at constant pressure (J/kg.K)
k_u = mixture thermal conductivity (J/m.K.s)
u_l = laminar burning velocity (m/s)
ρ_u = unburned mixture density (kg/m^3)
$R(c)$ = heat release rate source term (J/m^3.s)
λ = burning velocity eigenvalue
T_b = burned temperature (K)
T_u = unburned temperature (K)

Applying fractional mass weighting to equation 2 produces:

$$(QC_p/k_u)_b{}^{0.5}u_{lb}\rho_{ub} = x_i(QC_p/k_u)_i{}^{0.5}u_{li}\rho_{ui} + x_j(QC_p/k_u)_j{}^{0.5}u_{lj}\rho_{uj} \tag{4}$$

where
x_i = mass fraction of fuel *i*
x_j = mass fraction of fuel *j*
subscript b = fuel blend
subscript i = fuel *i*
subscript j = fuel *j*

Equation 4 will be used to calculate the laminar burning velocity of the fuel mixture in the present work. The unburned density, mean specific heat, thermal conductivity, burned temperature, and unburned temperature were calculated using GASEQ, a chemical equilibrium software for predicting combustion related reactions based on fuel compositions [13].

2. Research methodology

2.1 Combustion rig setup

The combustion rig in the current work is a constant volume mixing loop, manufactured from copper tubes, compression fittings, mixing fans, quartz tube of 0.65 m length, two pressure transducers, vacuum pump, valves, fuel injection port and an ignition port. Originally designed by Pennington [14], it was used to study the propagation speed of different gaseous fuels. The schematic diagram of the rig can be found in Figure 1.

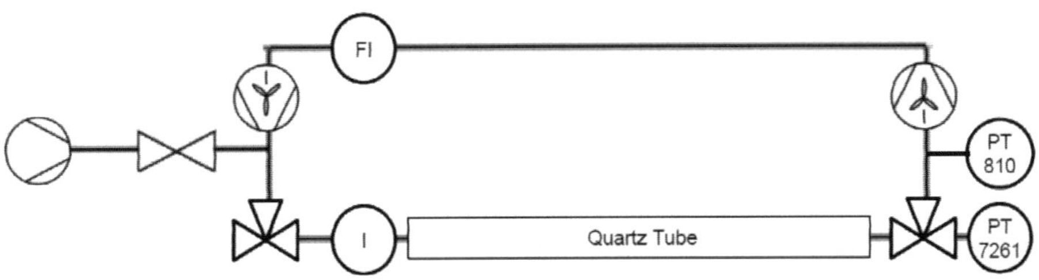

Figure 1. Schematic diagram of combustion rig consisting of valves, copper tubes, quartz tube, mixing fans, vacuum pump, fuel ignition port (I), fuel injection port (FI), and two pressure transducers.

Operating the rig was rather straightforward, the vacuum pump was used to remove air from the mixing loop, followed by injecting the appropriate amount of fuel via the fuel injection port, which was then mixed with air by opening either one of the three-way valves. Once air has been introduced into the mixing loop, the three-way valve was closed, and mixing fans were turned on for 3 minutes, to ensure that the fuel and air were properly mixed, producing a homogenous mixture. Both three-way valves were then opened, which isolates the quartz tube from the mixing loop, followed by igniting the mixture via the fuel ignition port using a pilot flame.

The flame propagation was recorded using a Phantom V210 High Speed Camera at a speed of 1500 frames per second (FPS) over a length of 0.505 m, calibrated before conducting the experiments. Each experiment was repeated three times to ensure repeatability of the experiments. All flames will have their burning velocities measured in the first 0.03 seconds of propagation after first appearing in the recorded part of the quartz tube, to minimize oscillatory effect in the middle part of the tube.

2.2 Fuel calculation

The usual method in calculating the fuel required would be by selecting the $\bar{x}_{fh}$ and ϕ of a specific fuel of interest to be applied to equation 1 to find the $\bar{x}_i$ and $\bar{x}_j$, which in the present work can be denoted as $\bar{x}_m$ and $\bar{x}_h$, since the fuel being mixed were methane and hydrogen. The present work took a different approach by choosing a constant theoretical laminar burning velocity u_l of 0.6 ms^{-1} for four different ϕ, which were 0.6, 0.8, 1.0 and 1.2, chosen based on the flammability limit of methane [15].

Utilizing the fifth blending law of equation 4, the corresponding value of $\bar{x}_{fh}$ for four mixtures were obtained with a theoretical u_l of 0.6 ms^{-1}. The theoretical u_l chart were plotted against $\bar{x}_{fh}$ in Figure 2, which consists of 4 different lines representing ϕ = 0.6, 0.8, 1.0 and 1.2. A horizontal line which represented u_l = 0.6 ms^{-1} was added to the chart to find the $\bar{x}_{fh}$ values for each ϕ at the desired u_l. Four $\bar{x}_{fh}$ values extracted from the intersections were 0.899 for ϕ = 0.6, 0.628 for ϕ = 0.8, 0.465 for ϕ = 1.0, and 0.425 for ϕ = 1.2.

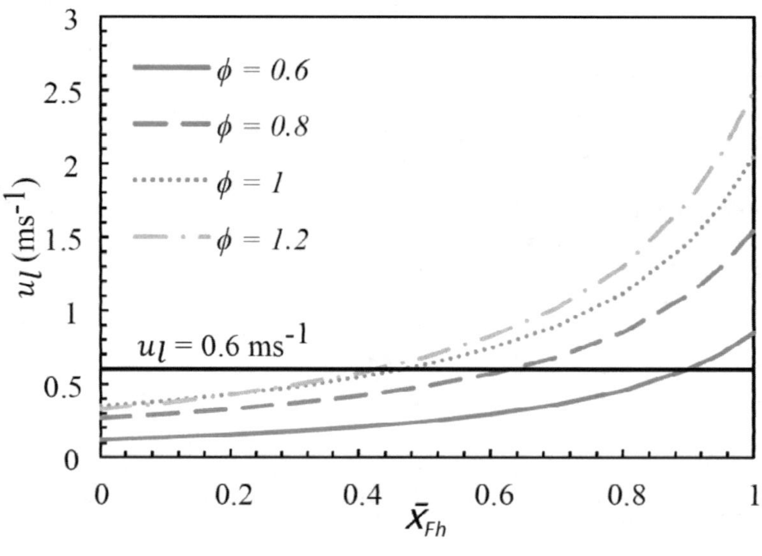

Figure 2. Theoretical $\boldsymbol{u_l}$ as function of $\bar{\boldsymbol{x}}_{\boldsymbol{fh}}$ for $\boldsymbol{\phi}$ = 0.6, 0.8, 1.0 and 1.2 using the fifth blending law.

The extracted values of $\bar{x}_{fh}$ were used to calculate the fractional mass weighting for both methane and hydrogen, $\bar{x}_m$ and $\bar{x}_h$ respectively. Calculate values were tabulated in table 2. Experiments will be conducted using these values, repeated 3 times for each mixture.

Table 2. Fractional mole ratios calculated at different ϕ values.

ϕ	$\bar{x}_{Fh}$	$\bar{x}_m$	$\bar{x}_h$	CH_4 (ml)	H_2 (ml)
0.6	0.899	0.276	0.724	19	171
0.8	0.628	0.658	0.342	60	101
1.0	0.465	0.782	0.218	87	76
1.2	0.425	0.802	0.198	105	78

3. Results and discussions

3.1 Flame tracking and visual comparison

Total of 12 flames were recorded and tracked using the native software of the Phantom V210 high-speed camera, Phantom Camera Control [16]. The tracked feature of the flames were their flame front position as performed by Ebieto *et al.* [17] and Gu *et al.* [18]. The tracked flames were then plotted against time to remove any flames which behaved differently from the others of the same mixture.

The ϕ = 0.6 flame appears really faint as shown in Figure 3 due to the high hydrogen content, $\bar{x}_{fh}$ = 0.899 and also the low equivalence ratio of 0.6, which indicates a lower concentration of methane compared to other mixtures tested. Hydrogen flames are inherently invisible in nature due to the absence of carbon atoms in their structure [19]. The brightness and contrast had to be adjusted to make the tracking process easier. The flame appeared to be convex in the initial stage, slightly compressed in the oscillatory stage and returned convex in the final stage.

Figure 3. Flame shape of ϕ = 0.6 flame (third run) during a) initial stage, b) oscillatory stage and c) final stage of propagation.

All three flames behaved similarly, starting with a smooth initial propagation, followed by an oscillatory period and returned to normal towards the end of the tube. The measured propagation speed in the initial period of propagation (0.03 seconds) for the three runs were 2.055 ms^{-1}, 2.516 ms^{-1} and 2.318 ms^{-1} respectively.

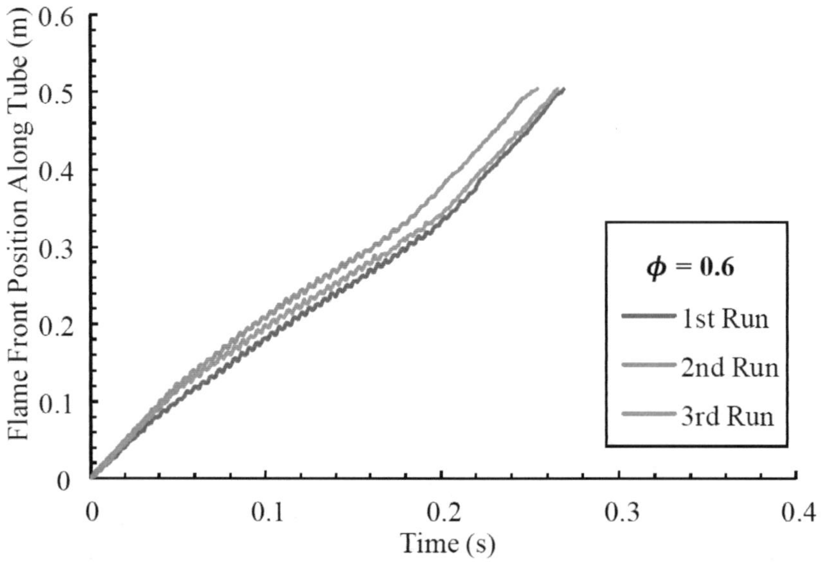

Figure 4. Tracked flame position for ϕ = 0.6 flames against time.

The ϕ = 0.8 flames were brighter compared to the ϕ = 0.6 flames, mainly due to the fact that the equivalence ratio was higher, and a lower hydrogen content, $\bar{x}_{fh}$ = 0.628, indicating a higher methane concentration in the former mixture. The initial flame shape appeared to be convex as shown in Figure 5, followed by a stretched flame shape in the oscillatory stage, and ended with a slightly asymmetric convex flame shape.

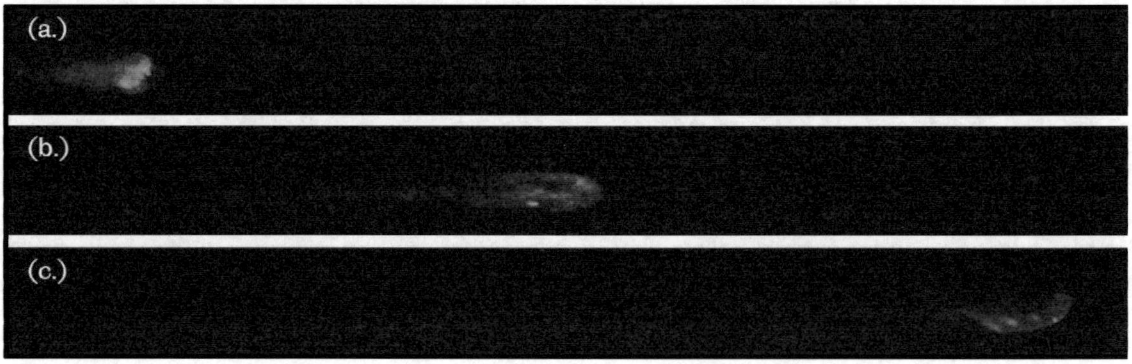

Figure 5. Flame shape of ϕ = 0.8 flame (first run) during a) initial stage, b) oscillatory stage and c) final stage of propagation.

The asymmetric shape observed in the final stage was attributed to hot air rising, causing it to tip towards the right side, as observed by Yang [20] in the same rig and Guénoche [21] in an upward vertical flame rig. It was also observed that the tipping led to a higher propagation speed towards the end of the first run, shown in Figure 6. However, this run was not removed from being analysed further since it has the same initial propagation with the second run, measured to be 1.791 ms^{-1} and 2.002 ms^{-1} respectively. The third run was removed from further analysis due to its slower propagation behaviour in the first 0.03 seconds.

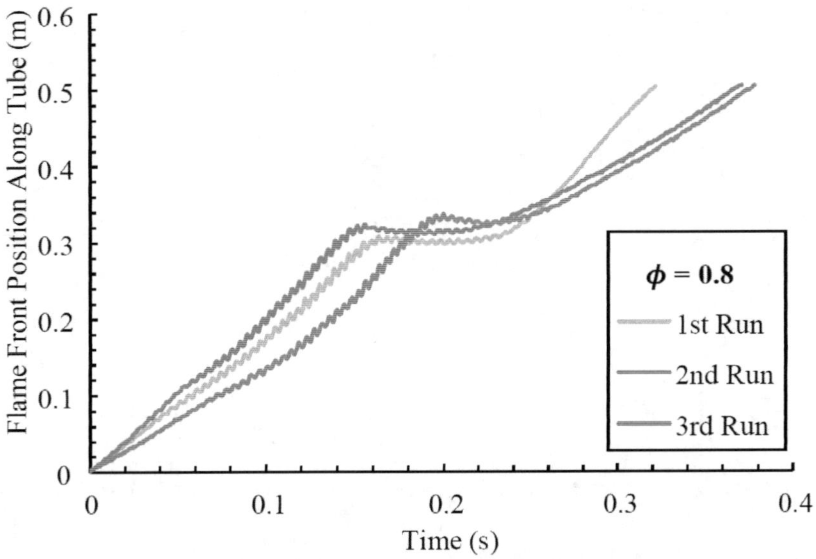

Figure 6. Tracked flame position for ϕ = 0.8 flames against time.

As expected, the ϕ = 1.0 flames were brighter compared to the previous flames discussed for the same reason, a higher methane concentration. The initial flame shape appeared to be convex as shown in Figure 7 followed by a stretched flame shape in the oscillatory stage, and returned to its initial convex flame shape. One new addition to this particular mixture was the existence of an elongated flame tip in the final stage, which was not previously observed by Yang *et al.*[20] and Ebieto *et al.* [22], who did not adapt the blending law used in the present work.

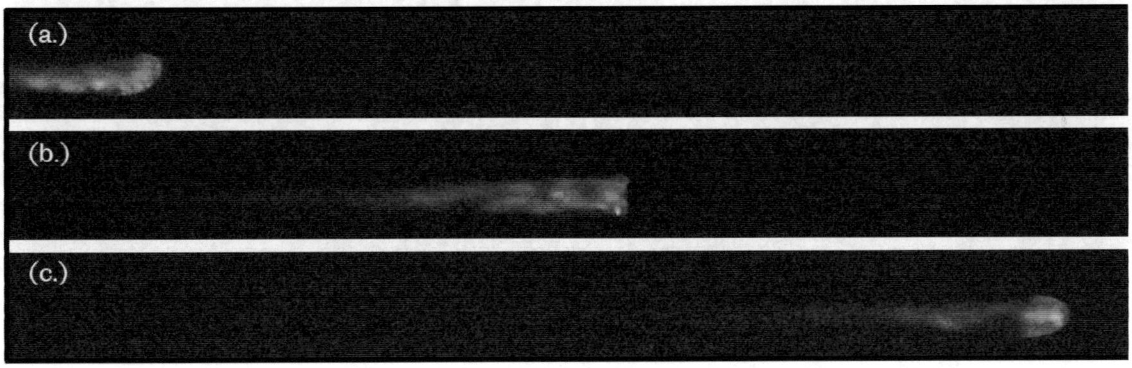

Figure 7. Flame shape of ϕ = 1.0 flame (third run) during a) initial stage, b) oscillatory stage and c) final stage of propagation.

The elongated tail was expected to cause the third flame to have a faster propagation time, which was present in both the second and third run. A longer tail indicates a higher surface area of the flame, which is directly proportional to the burning rate of a flame [23]. The first run was excluded from further analysis due to its slower propagation rate compare to other runs as shown in Figure 8. Their initial propagation speeds were 2.463 ms^{-1} and 2.661 ms^{-1} respectively.

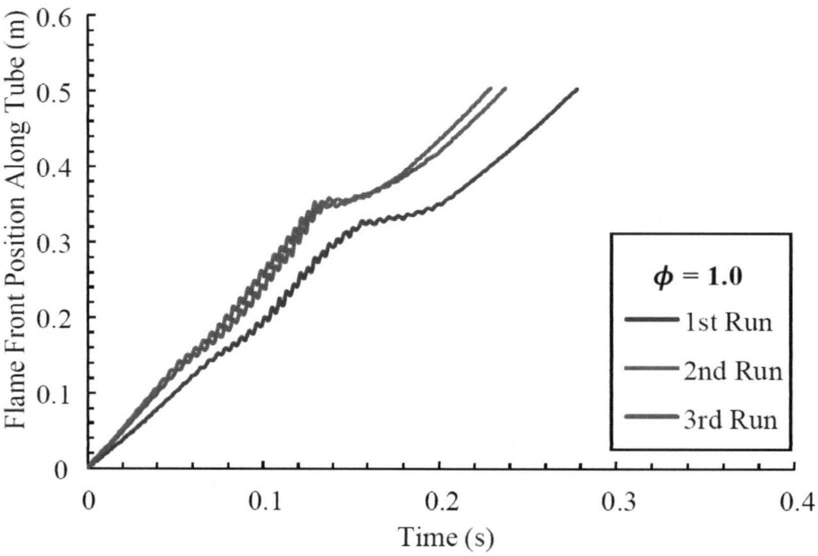

Figure 8. Tracked flame position for ϕ = 1.0 flames against time.

Tracking the ϕ = 1.2 flames were easy due to their brightness. The flames appeared symmetric in both the oscillatory and final stage of propagation as shown in Figure 9. The initial flame shape was slightly tipped as observed in the ϕ = 0.8 flame in Figure 5, but appeared in the initial stage instead of the final stage. The elongated flame tail present in the ϕ = 1.2 flame was also present in this flame.

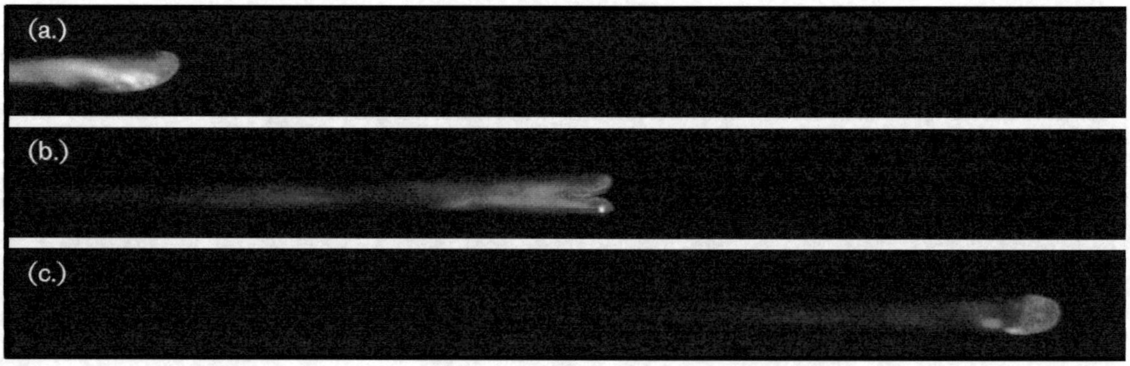

Figure 9. Flame shape of ϕ = 1.2 flame (third run) during a) initial stage, b) oscillatory stage and c) final stage of propagation.

The ϕ = 1.2 flames were the only rich flames tested in the present work. The higher concentration of methane appeared to produce fastest propagating flames compared to other mixtures, which was expected. Despite having the lowest hydrogen concentration of $\bar{x}_{fh}$ = 0.425, they managed to propagate completely before 0.22 seconds. This can be explained by the shift in peak flame speed towards the richer region for hydrogen observed by Hu *et al.* [24], which adapted the second blending law, the fractional mole weighting blending law. Referring to Figure 10, it was obvious that the first run was excluded, since it has a significantly higher initial propagation compared to the other two runs. The flame speed for the second and third run were measured to be 2.845 ms^{-1} and 2.700 ms^{-1}.

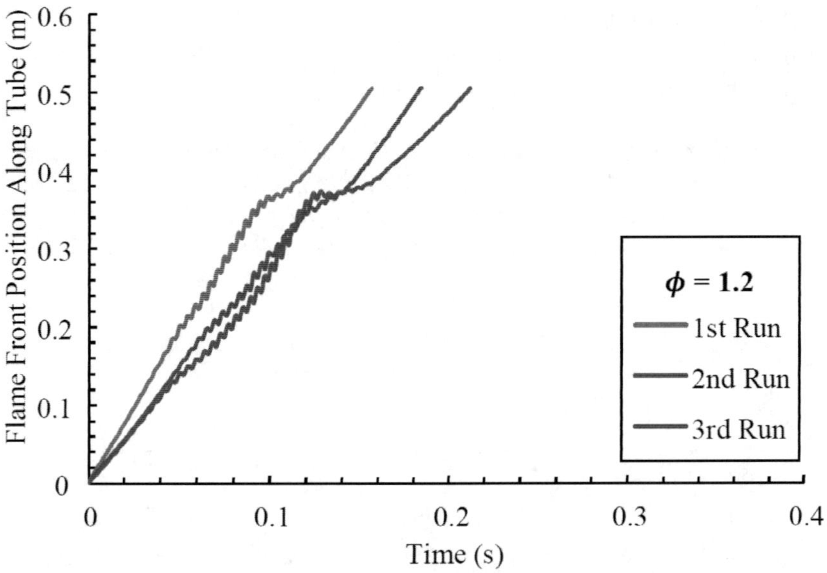

Figure 10. Tracked flame position for ϕ = 1.2 flames against time.

3.2 Measurement of laminar burning velocity

The measured flame speed from the previous section was used to calculate the laminar burning velocity. Initial flame front positions were compiled in Figure 11, which were used to calculate the flame speed in the previous section.

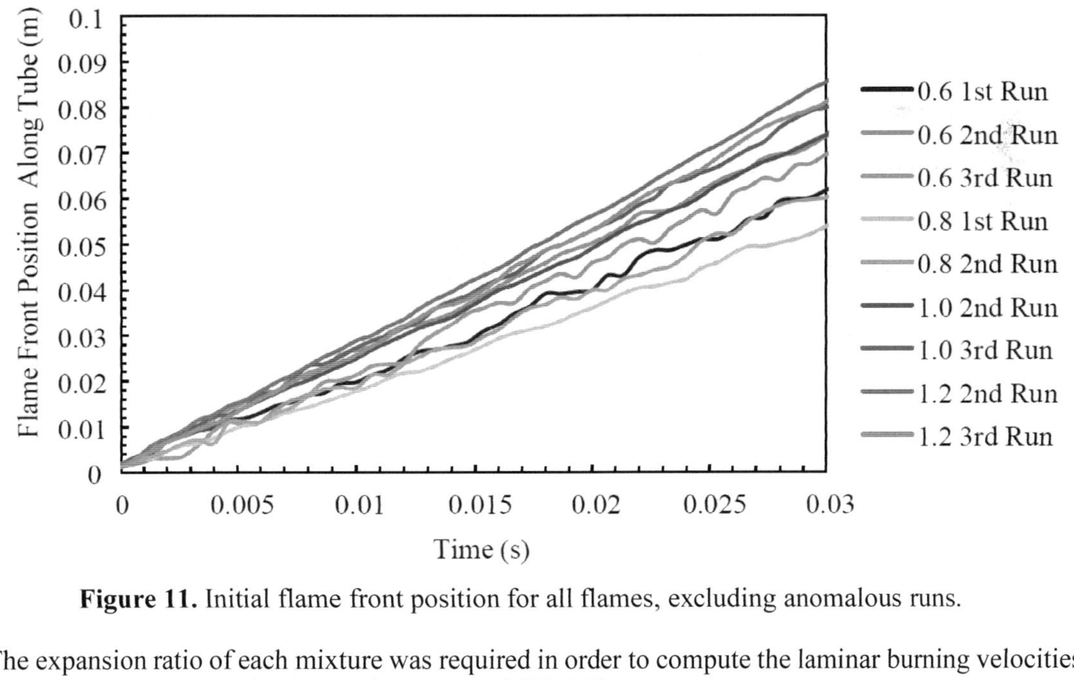

Figure 11. Initial flame front position for all flames, excluding anomalous runs.

The expansion ratio of each mixture was required in order to compute the laminar burning velocities, which was a common practice in previous research[18, 24]:

$$u_l = S_l \frac{\rho_b}{\rho_u} \tag{5}$$

where S_l = measured flame speed (m.s^{-1})
ρ_b = burned mixture density (kg.m^{-3})
ρ_u = unburned mixture density (kg.m^{-3})

Calculated values were then tabulated in table 3. Comparing them to the predicted laminar burning velocity of 0.6 ms^{-1}, the measured values were significantly lower. The average of all runs was calculated to be 0.368 ms^{-1} with a standard deviation of 0.056 ms^{-1}. It was observed that the ϕ = 0.6 flames had the closest value to the predicted value, which was 0.419 ms^{-1}, followed by ϕ = 1.2, ϕ = 1.0 and finally ϕ = 0.8.

Table 3. Fractional mole ratios calculated at different ϕ values.

ϕ	$\frac{\rho_b}{\rho_u}$	S_l (ms^{-1})			u_l (ms^{-1})			
		1	2	3	1	2	3	Avg.
0.6	0.183	2.055	2.516	2.318	0.375	0.459	0.423	0.419
0.8	0.154	1.791	2.002	N/A	0.276	0.308	N/A	0.292
1.0	0.138	N/A	2.463	2.661	N/A	0.339	0.366	0.352
1.2	0.138	N/A	2.845	2.700	N/A	0.394	0.374	0.384

The deviation observed in the experimental u_l values from the predicted values showed that other significant factors were in play. One of the main factors were thought to be the pressure pulsations which were inherent in flame tube rigs, caused by acoustic interactions within the tube, causing the oscillatory behavior in the middle of the tube. The initial propagation may appear to be steady, but low-pressure oscillations may have started forming, leading to the low experimental values.

Apart from low pressure oscillations, the flame tipping observed may have also affected the overall flame propagation. Flame tipping was thought to be caused by the density difference between the burned and unburned gasses, causing the flame to propagate in an asymmetric manner. Finally, the heat loss to the surrounding was thought to contribute to the slow flame speed, which is difficult to avoid with flame tube rigs.

4. Conclusion

In an attempt to produce similar flame propagation from mixtures with similar theoretical u_l of 0.6 ms^{-1} using the fifth fuel blending law proposed by Bradley *et al.* [9], it was discovered that the flames propagated with increasing speed from ϕ = 0.8 – 1.2 (0.292 ms^{-1}, 0.352 ms^{-1}, and 0.384 ms^{-1} respectively), with ϕ = 0.6 being an exception, which propagated the fastest among the mixtures. This was attributed to the high hydrogen content, $\bar{x}_{fh}$ = 0.899, which translates to 90% mass percentage hydrogen of the binary fuel. Apart from the speed difference, the flame oscillatory behaviour was also noticeably different, where most of them were oscillated significantly (ϕ = 0.8 – 1.2), while ϕ = 0.6 flames were only slightly vibrating. The blending law used may produce mixtures with similar theoretical laminar burning velocities, but it does not promise similar combustion behaviour, leading to different experimental laminar burning velocities.

References

[1] Çeper B A 2012 Use of Hydrogen-Methane Blends in Internal Combustion Engines. *Hydrogen Energy - Challenges and Perspectives* Minic D (ed) (London: InTechOpen)

[2] Ma F, Wang Y, Liu H, Li Y, Wang J and Zhao S 2007 Experimental study on thermal efficiency and emission characteristics of a lean burn hydrogen enriched natural gas engine *Int. J. Hydr. Ener.* **32** 5067–75

[3] Cho H M and He B Q 2007 Spark ignition natural gas engines-A review *Energy Conversion Management* **48** 608–18

[4] Pal A and Agarwal A K 2015 Comparative study of laser ignition and conventional electrical spark ignition systems in a hydrogen fuelled engine *Int. J. Hydr. Ener.* **40** 2386–95
[5] Van Lipzig J P J, Nilsson E J K, De Goey L P H nd Konnov A A 2011 Laminar burning velocities of n-heptane, iso-octane, ethanol and their binary and tertiary mixtures *Fuel* **90** 2773–81
[6] Payman W and Wheeler R V 1922 The composition of gaseous fuels in relation to their utilisation *Fuel in Science and Practice* **1** 185–96
[7] Di Sarli V and Di Benedetto A 2007 Laminar burning velocity of hydrogen-methane/air premixed flames *Int. J. of Hydr. Ener.* **32** 637–46.
[8] Bradley D, Habik S E D and El-Sherif S A 1991 A generalization of laminar burning velocities and volumetric heat release rates *Combustion and Flame* **87** 336–45
[9] Bradley D, Lawes M and Mumby R 2017 Burning velocity and Markstein length blending laws for methane/air and hydrogen/air blends *Fuel* **187** 268–75
[10] Hirasawa T, Sung C J, Joshi A, Yang Z, Wang H and Law CK 2002 Determination of laminar flame speeds using digital particle image velocimetry: Binary fuel blends of ethylene, n-butane, and toluene *Proceedings of the Combustion Institute* **29** 1427–34
[11] Spalding D B 1957 I. Predicting the laminar flame speed in gases with temperature-explicit reaction rates *Combustion and Flame* **1** 287–295.
[12] Spalding D B 1957 II. One-dimensional laminar flame theory for temperature-explicit reaction rates *Combustion and Flame* **1** 296–307
[13] Morley C 2005 *GASEQ* Version 0.79 (www.gaseq.co.uk: Morley C)
[14] Pennington P *Flames in Tubes[master's thesis]* 2009 (Sheffield: University of Sheffield)
[15] Amaludin N A 2020 *Flame Instabilities in Open Ended Tubes [dissertation]* (Sheffield: University of Sheffield)
[16] Vision Research 2016 *Phantom Camera Control* Version 2.7 (New Jersey: Vision Research)
[17] Ebieto C E 2017 *Dynamics of Premixed Flames in Tube[dissertation]* (Sheffield: University of Sheffield)
[18] Gu X J, Haq MZ, Lawes M and Woolley R 2000 Laminar burning velocity and Markstein lengths of methane–air mixtures *Combustion and Flame* **121** 41–58
[19] Glassman I and Yetter R A 2008 *Combustion* 4th Edition (Cambridge: Academic Press)
[20] Yang J, Mossa F M S, Huang H W, Wang Q, Woolley R and Zhang Y 2015 Oscillating flames in open tubes *Proceedings of the Combustion Institute* (The Combustion Institute) 2075–2082.
[21] Guénoche H Flame Propagation in Tubes and in Closed Vessels in: *Nonsteady Flame Propagation* 107–181
[22] Ebieto C E, Amaludin N A and Woolley R 2015 Methane / Hydrogen / Air Flame Oscillations in Open Ended Tubes *Proceedings of the 25th International Colloquium on the Dynamics of Explosion and Reactive Systems* (The Institute for the Dynamics of Explosions and Reactive Systems)
[23] Markstein G H 1964 *Nonsteady Flame Propagation* (New York: Pergamon Press)
[24] Hu E, Huang Z, He J and Jin C, Zheng J 2009 Experimental and numerical study on laminar burning characteristics of premixed methane-hydrogen-air flames.*Int. J. of Hydrogen Energy* **34** 4876–88

IOP Conf. Series: Materials Science and Engineering 1217 (2022) 012013 doi:10.1088/1757-899X/1217/1/012013

Methane hydrogen laminar burning velocity blending laws in horizontal open-ended flame tube rig

N A Amaludin[1*], M Morrow[2], R Woolley[2], A E Amaludin[3*]

[1]Centre of Research in Energy and Advanced Materials, Faculty of Engineering, Universiti Malaysia Sabah, Kota Kinabalu, 88400, Sabah, Malaysia.
[2]Mechanical Engineering Department, Engineering Heartspace, Sir Frederick Mappin Building, Mappin Street, Sheffield S1 3JD, United Kingdom.
[3]Faculty of Engineering, Universiti Malaysia Sabah, Kota Kinabalu, 88400, Sabah, Malaysia.

[*]Corresponding author: nazrein.amaludin@ums.edu.my, adriana@ums.edu.my

Abstract. Different fuel properties and chemical kinetics of two different fuels would make it challenging to predict the combustion parameters of a binary fuel. Understanding the effect of blending methane and hydrogen gas is the main focus of this paper. Utilizing a horizontal tube combustion rig, methane-hydrogen fuel blends were created using blending laws from past literature, over a range of equivalence ratios from 0.6 – 1.2 were studied, while keeping one combustion parameter constant, the theoretical laminar burning velocity. The selected theoretical laminar burning velocity for all the mixtures tested were kept constant at 0.6 ms^{-1}. Different factors affected the flame propagation across the tube, including acoustic pressure oscillations, heat loss from the rig, and obvious difference in hydrogen percentage in the fuel blends. The average experimental laminar burning velocity of all the flames was 0.368 ms^{-1}, compared to the expected value of 0.6 ms^{-1}. In an attempt to keep the theoretical laminar burning velocity constant for different mixtures, it was discovered that this did not promise the same flame propagation behaviour for the tested mixtures. Further experimentation and analysis are required in order to better understand the underlying interaction of the fuel blends.

1. Introduction

Carbon emissions from fossil fuels have been increasing at an alarming rate, which initiated the interest in using alternative fuels to reduce the global carbon footprint. An extensive research and development would be required to design machines that operate on alternative fuels alone due to the obvious difference between fossil fuels and alternative fuels [1]. Apart from that, migration to alternative fuels would leave the current machineries to be left unused, causing further loss to existing fossil fuel users. In order to make the migration to alternative fuels less intrusive, mixing fossil fuels and alternative fuels seems to be the best option, but it will not be an easy task. The different fuel properties and chemical kinetics of the fuels would make it challenging to predict the combustion parameters of the binary fuel [2]. Methane is a preferred hydrocarbon to mix with alternative fuels since it is the cleanest burning fossil fuel, due to its simple structure [3]. On the other hand, hydrogen is an alternative fuel which has one of the highest potential, thanks to its high burning rate and carbon free structure [4].

Content from this work may be used under the terms of the Creative Commons Attribution 3.0 licence. Any further distribution of this work must maintain attribution to the author(s) and the title of the work, journal citation and DOI.
Published under licence by IOP Publishing Ltd

1.1 Fuel composition

The factors which determine the fuel composition of a binary fuel are the equivalence ratio, ϕ, mole fraction of the first fuel, $\bar{x}_{fi}$, and the mole fraction of the second fuel, $\bar{x}_{fj}$. It is important to note that different fuels require different amount of air in order to oxidize completely, thus the need to include the air required by both fuels in the following manner:

$$\bar{x}_{fi} = \frac{\left[{\phi}/{(\phi + a_{si})}\right]\bar{x}_i}{\left[{\phi}/{(\phi + a_{si})}\right]\bar{x}_i + \left[{\phi}/{(\phi + a_{sj})}\right]\bar{x}_j} \tag{1}$$

where $\bar{x}_f$ = mole fraction of fuel within the binary fuel mixture.
ϕ = equivalence ratio.
a_s = stoichiometric air mole/fuel mole.
$\bar{x}$ = mole fraction fuel within the air-fuel mixture, $\bar{x}_i + \bar{x}_j = 1$
subscript *i* = fuel *i*
subscript *j* = fuel *j*

The normal approach in utilizing equation 1 is by setting value for $\bar{x}_{fi}$ and ϕ to obtain an equation with two unknowns, $\bar{x}_i$ and $\bar{x}_j$. The equation will then be solved using equation $\bar{x}_i + \bar{x}_j = 1$ to obtain the value of $\bar{x}_i$ and $\bar{x}_j$. Fractional mass weightings are calculated by multiplying the mole fractions with the molecular weight of the respective fuels, followed by normalizing the mass weightings to 1. The mole and mass fraction of fuels within the air-fuel mixture will then be used in laminar burning velocity blending laws.

1.2 Laminar burning velocity blending laws

Six different laminar burning velocity blending (u_l) laws were commonly used over the years, which were based on fractional mass weighting of u_l of constituent mixtures law [5], fractional mole weighting of u_l of constituent mixtures [6], modified Le Châtelier law [7], laminar burning velocity plotted against molar heat of reaction ($\bar{Q}$) per mole of constituent mixture [8], fractional mass weighting of constituent mixture $u_l\rho_u$ with the expression of mass rate of burning [9], and finally fractional mole weighting of the activation temperature, $\tilde{T}_a$, used to calculate the $\tilde{T}_a$ of the blend. Details of the blending laws were tabulated in table 1, which shows that the fifth law has the smallest ratio of predicted laminar burning velocity to the actual laminar burning velocity, at 1.01, with a standard deviation of 0.03.

Table 1. Different laminar burning velocity blending laws for binary fuels.

No.	*Laminar Burning Velocity Blending Law*	Ref.	u_{lp}/u_l	σ
1	Fractional mass weighting of u_l of constituent mixtures.	[5]	1.09	0.06
2	Fractional mole weighting of u_l of constituent mixtures.	[6]	1.2	0.09
3	Modified Le Châtelier law. Fractional mole weighting of constituent fuels.	[7]	0.92	0.04
4	u_l plotted against $\bar{Q}$ per mole of constituent mixture.	[8]	1.18	0.08
5	Fractional mass weighting of constituent mixture $u_l\rho_u$ with $(QC_p/k_u)^{0.5}$.	[9]	1.01	0.03
6	Fractional mole weighting of $\tilde{T}_a$ to obtain blend $\tilde{T}_a$.	[10]	0.89	0.04

The current work utilized the fifth law proposed by Bradley *et al.* [9], since it has the best prediction to experimental velocity ratio of 1.01, and standard deviation of 0.03. This law was based on the work of Spalding [11, 12], where the mass of rate of burning was rearranged in the following form:

$$(Qc_p/k_u)^{0.5}u_l\rho_u = \left[\frac{\int_0^1 R(c)dc}{\lambda}\right]^{0.5} \tag{2}$$

$$Q = c_p(T_b - T_u) \tag{3}$$

where
- Q = mass heat of reaction (J/kg)
- c_p = mean specific heat at constant pressure (J/kg.K)
- k_u = mixture thermal conductivity (J/m.K.s)
- u_l = laminar burning velocity (m/s)
- ρ_u = unburned mixture density (kg/m^3)
- $R(c)$ = heat release rate source term (J/m^3.s)
- λ = burning velocity eigenvalue
- T_b = burned temperature (K)
- T_u = unburned temperature (K)

Applying fractional mass weighting to equation 2 produces:

$$(QC_p/k_u)_b{}^{0.5}u_{lb}\rho_{ub} = x_i(QC_p/k_u)_i{}^{0.5}u_{li}\rho_{ui} + x_j(QC_p/k_u)_j{}^{0.5}u_{lj}\rho_{uj} \tag{4}$$

where
- x_i = mass fraction of fuel *i*
- x_j = mass fraction of fuel *j*
- subscript b = fuel blend
- subscript i = fuel *i*
- subscript j = fuel *j*

Equation 4 will be used to calculate the laminar burning velocity of the fuel mixture in the present work. The unburned density, mean specific heat, thermal conductivity, burned temperature, and unburned temperature were calculated using GASEQ, a chemical equilibrium software for predicting combustion related reactions based on fuel compositions [13].

2. Research methodology

2.1 Combustion rig setup

The combustion rig in the current work is a constant volume mixing loop, manufactured from copper tubes, compression fittings, mixing fans, quartz tube of 0.65 m length, two pressure transducers, vacuum pump, valves, fuel injection port and an ignition port. Originally designed by Pennington [14], it was used to study the propagation speed of different gaseous fuels. The schematic diagram of the rig can be found in Figure 1.

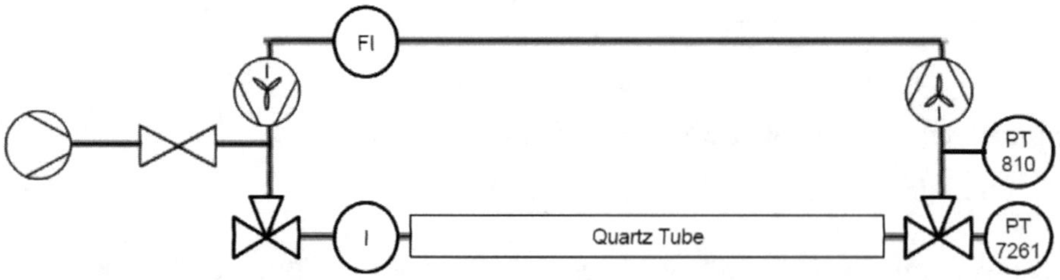

Figure 1. Schematic diagram of combustion rig consisting of valves, copper tubes, quartz tube, mixing fans, vacuum pump, fuel ignition port (I), fuel injection port (FI), and two pressure transducers.

Operating the rig was rather straightforward, the vacuum pump was used to remove air from the mixing loop, followed by injecting the appropriate amount of fuel via the fuel injection port, which was then mixed with air by opening either one of the three-way valves. Once air has been introduced into the mixing loop, the three-way valve was closed, and mixing fans were turned on for 3 minutes, to ensure that the fuel and air were properly mixed, producing a homogenous mixture. Both three-way valves were then opened, which isolates the quartz tube from the mixing loop, followed by igniting the mixture via the fuel ignition port using a pilot flame.

The flame propagation was recorded using a Phantom V210 High Speed Camera at a speed of 1500 frames per second (FPS) over a length of 0.505 m, calibrated before conducting the experiments. Each experiment was repeated three times to ensure repeatability of the experiments. All flames will have their burning velocities measured in the first 0.03 seconds of propagation after first appearing in the recorded part of the quartz tube, to minimize oscillatory effect in the middle part of the tube.

2.2 *Fuel calculation*

The usual method in calculating the fuel required would be by selecting the $\bar{x}_{fh}$ and ϕ of a specific fuel of interest to be applied to equation 1 to find the $\bar{x}_i$ and $\bar{x}_j$, which in the present work can be denoted as $\bar{x}_m$ and $\bar{x}_h$, since the fuel being mixed were methane and hydrogen. The present work took a different approach by choosing a constant theoretical laminar burning velocity u_l of 0.6 ms^{-1} for four different ϕ, which were 0.6, 0.8, 1.0 and 1.2, chosen based on the flammability limit of methane [15].

Utilizing the fifth blending law of equation 4, the corresponding value of $\bar{x}_{fh}$ for four mixtures were obtained with a theoretical u_l of 0.6 ms^{-1}. The theoretical u_l chart were plotted against $\bar{x}_{fh}$ in Figure 2, which consists of 4 different lines representing ϕ = 0.6, 0.8, 1.0 and 1.2. A horizontal line which represented u_l = 0.6 ms^{-1} was added to the chart to find the $\bar{x}_{fh}$ values for each ϕ at the desired u_l. Four $\bar{x}_{fh}$ values extracted from the intersections were 0.899 for ϕ = 0.6, 0.628 for ϕ = 0.8, 0.465 for ϕ = 1.0, and 0.425 for ϕ = 1.2.

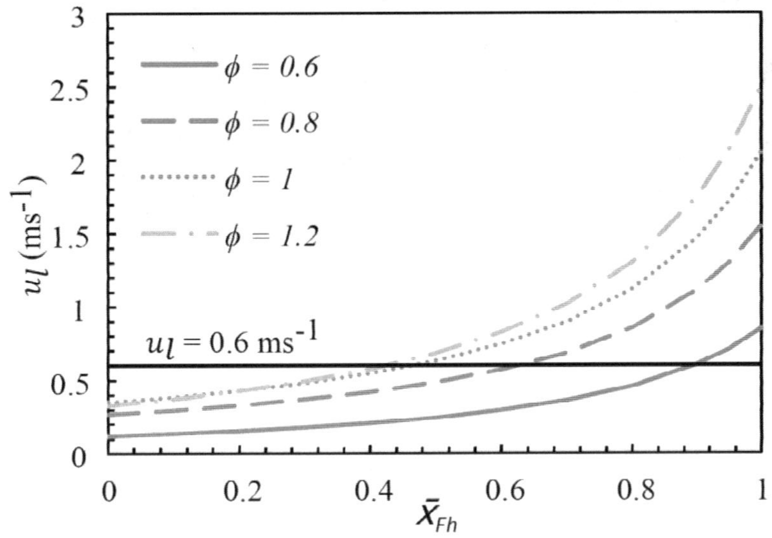

Figure 2. Theoretical $\boldsymbol{u_l}$ as function of $\bar{\boldsymbol{x}}_{\boldsymbol{fh}}$ for $\boldsymbol{\phi}$ = 0.6, 0.8, 1.0 and 1.2 using the fifth blending law.

The extracted values of $\bar{x}_{fh}$ were used to calculate the fractional mass weighting for both methane and hydrogen, $\bar{x}_m$ and $\bar{x}_h$ respectively. Calculate values were tabulated in table 2. Experiments will be conducted using these values, repeated 3 times for each mixture.

Table 2. Fractional mole ratios calculated at different ϕ values.

ϕ	$\bar{x}_{Fh}$	$\bar{x}_{m}$	$\bar{x}_{h}$	CH_4 (ml)	H_2 (ml)
0.6	0.899	0.276	0.724	19	171
0.8	0.628	0.658	0.342	60	101
1.0	0.465	0.782	0.218	87	76
1.2	0.425	0.802	0.198	105	78

3. Results and discussions

3.1 Flame tracking and visual comparison

Total of 12 flames were recorded and tracked using the native software of the Phantom V210 high-speed camera, Phantom Camera Control [16]. The tracked feature of the flames were their flame front position as performed by Ebieto *et al.* [17] and Gu *et al.* [18]. The tracked flames were then plotted against time to remove any flames which behaved differently from the others of the same mixture.

The ϕ = 0.6 flame appears really faint as shown in Figure 3 due to the high hydrogen content, $\bar{x}_{fh}$ = 0.899 and also the low equivalence ratio of 0.6, which indicates a lower concentration of methane compared to other mixtures tested. Hydrogen flames are inherently invisible in nature due to the absence of carbon atoms in their structure [19]. The brightness and contrast had to be adjusted to make the tracking process easier. The flame appeared to be convex in the initial stage, slightly compressed in the oscillatory stage and returned convex in the final stage.

Figure 3. Flame shape of ϕ = 0.6 flame (third run) during a) initial stage, b) oscillatory stage and c) final stage of propagation.

All three flames behaved similarly, starting with a smooth initial propagation, followed by an oscillatory period and returned to normal towards the end of the tube. The measured propagation speed in the initial period of propagation (0.03 seconds) for the three runs were 2.055 ms^{-1}, 2.516 ms^{-1} and 2.318 ms^{-1} respectively.

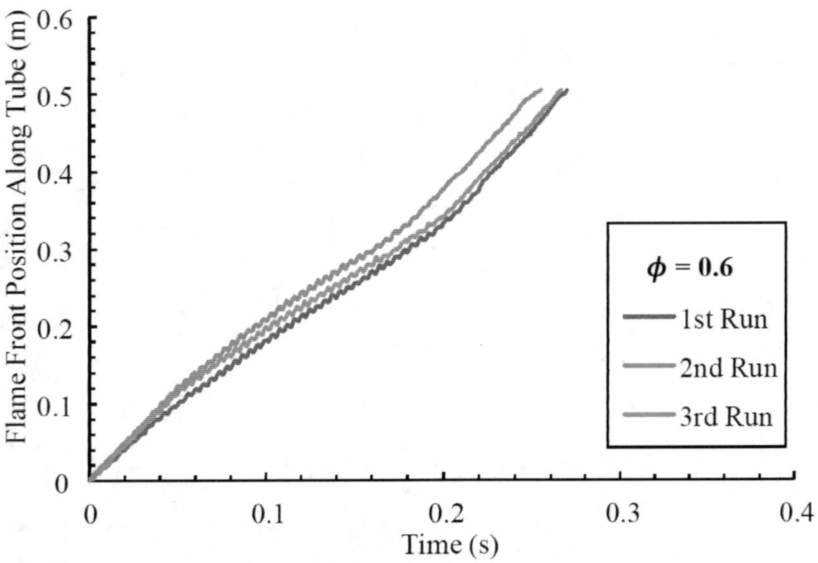

Figure 4. Tracked flame position for ϕ = 0.6 flames against time.

The ϕ = 0.8 flames were brighter compared to the ϕ = 0.6 flames, mainly due to the fact that the equivalence ratio was higher, and a lower hydrogen content, $\bar{x}_{fh}$ = 0.628, indicating a higher methane concentration in the former mixture. The initial flame shape appeared to be convex as shown in Figure 5, followed by a stretched flame shape in the oscillatory stage, and ended with a slightly asymmetric convex flame shape.

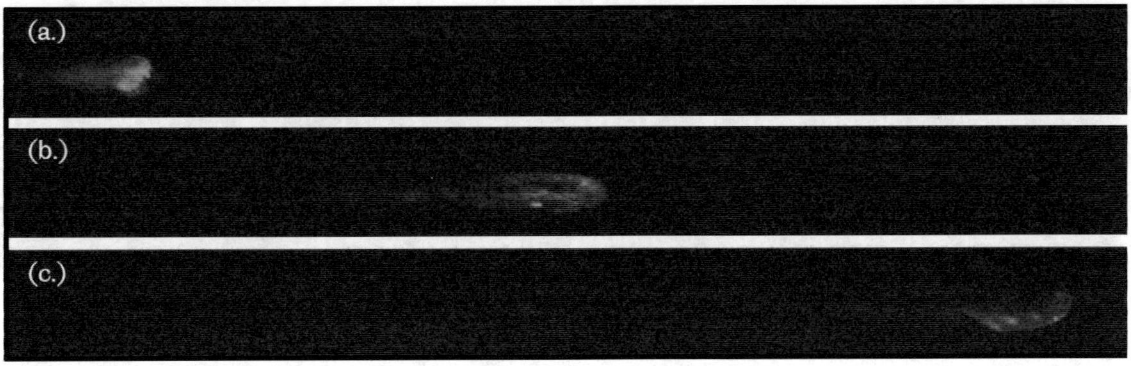

Figure 5. Flame shape of ϕ = 0.8 flame (first run) during a) initial stage, b) oscillatory stage and c) final stage of propagation.

The asymmetric shape observed in the final stage was attributed to hot air rising, causing it to tip towards the right side, as observed by Yang [20] in the same rig and Guénoche [21] in an upward vertical flame rig. It was also observed that the tipping led to a higher propagation speed towards the end of the first run, shown in Figure 6. However, this run was not removed from being analysed further since it has the same initial propagation with the second run, measured to be 1.791 ms^{-1} and 2.002 ms^{-1} respectively. The third run was removed from further analysis due to its slower propagation behaviour in the first 0.03 seconds.

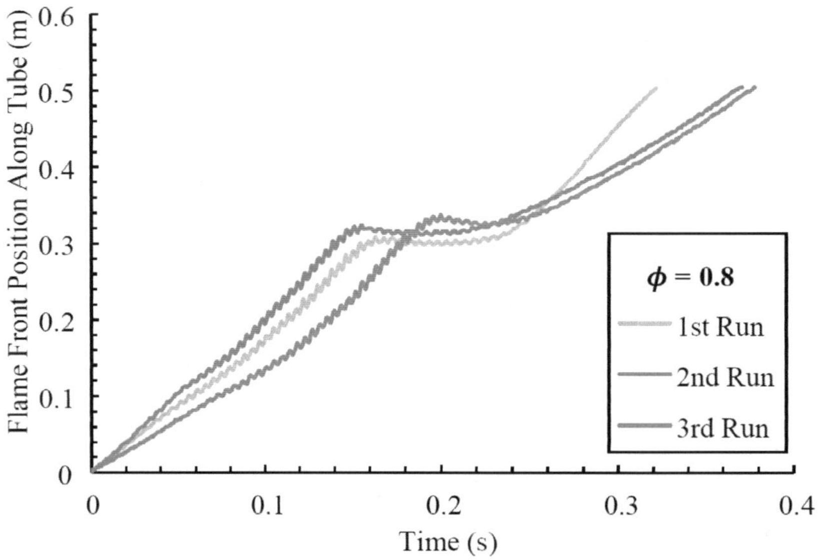

Figure 6. Tracked flame position for ϕ = 0.8 flames against time.

As expected, the ϕ = 1.0 flames were brighter compared to the previous flames discussed for the same reason, a higher methane concentration. The initial flame shape appeared to be convex as shown in Figure 7 followed by a stretched flame shape in the oscillatory stage, and returned to its initial convex flame shape. One new addition to this particular mixture was the existence of an elongated flame tip in the final stage, which was not previously observed by Yang *et al.*[20] and Ebieto *et al.* [22], who did not adapt the blending law used in the present work.

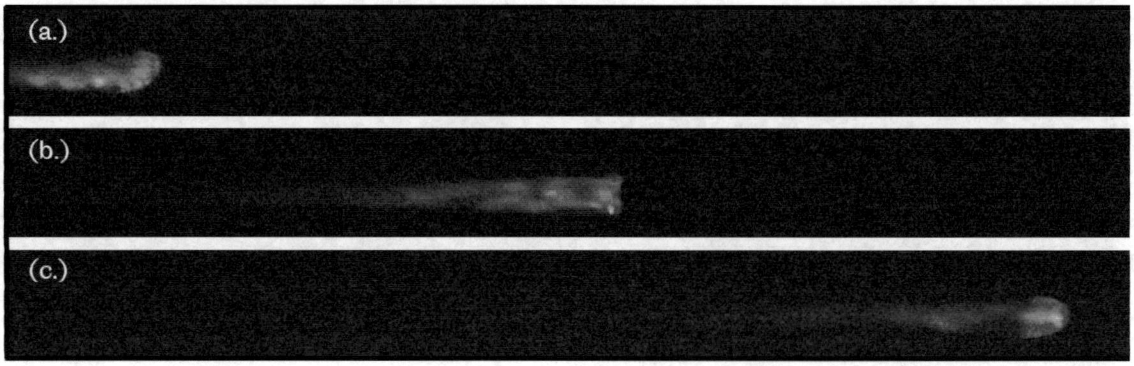

Figure 7. Flame shape of ϕ = 1.0 flame (third run) during a) initial stage, b) oscillatory stage and c) final stage of propagation.

The elongated tail was expected to cause the third flame to have a faster propagation time, which was present in both the second and third run. A longer tail indicates a higher surface area of the flame, which is directly proportional to the burning rate of a flame [23]. The first run was excluded from further analysis due to its slower propagation rate compare to other runs as shown in Figure 8. Their initial propagation speeds were 2.463 ms^{-1} and 2.661 ms^{-1} respectively.

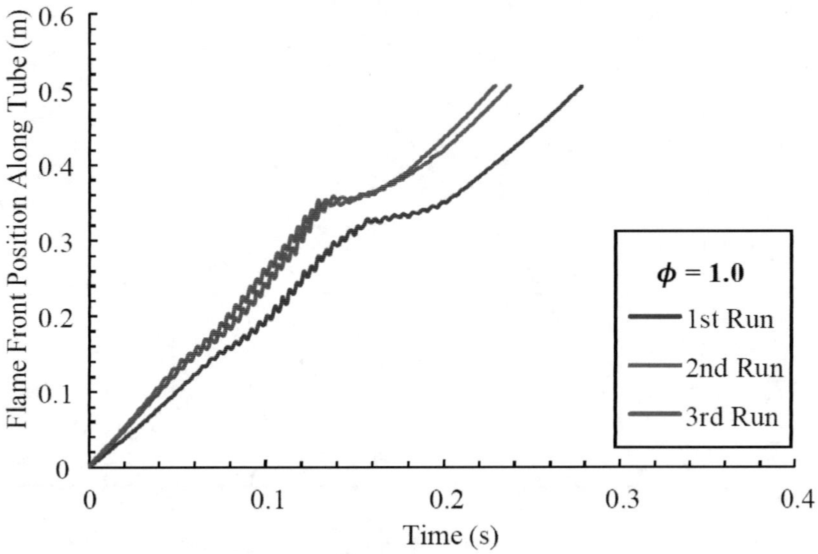

Figure 8. Tracked flame position for ϕ = 1.0 flames against time.

Tracking the ϕ = 1.2 flames were easy due to their brightness. The flames appeared symmetric in both the oscillatory and final stage of propagation as shown in Figure 9. The initial flame shape was slightly tipped as observed in the ϕ = 0.8 flame in Figure 5, but appeared in the initial stage instead of the final stage. The elongated flame tail present in the ϕ = 1.2 flame was also present in this flame.

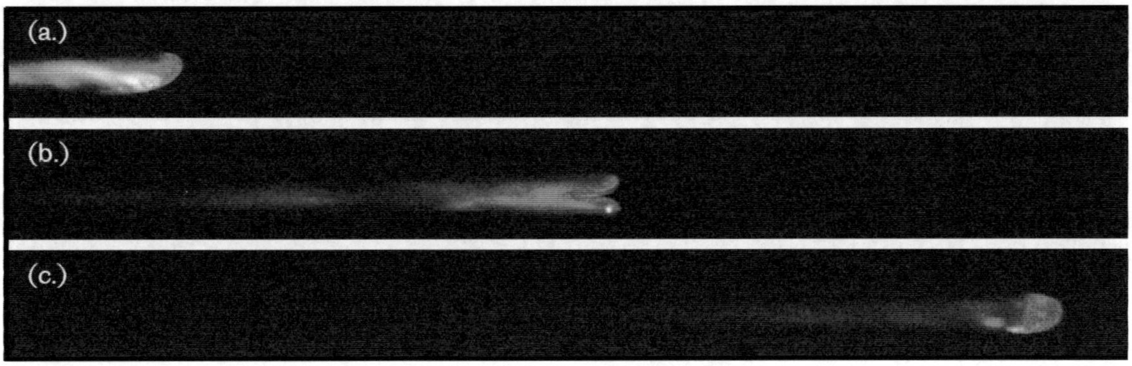

Figure 9. Flame shape of ϕ = 1.2 flame (third run) during a) initial stage, b) oscillatory stage and c) final stage of propagation.

The ϕ = 1.2 flames were the only rich flames tested in the present work. The higher concentration of methane appeared to produce fastest propagating flames compared to other mixtures, which was expected. Despite having the lowest hydrogen concentration of $\bar{x}_{fh}$ = 0.425, they managed to propagate completely before 0.22 seconds. This can be explained by the shift in peak flame speed towards the richer region for hydrogen observed by Hu *et al.* [24], which adapted the second blending law, the fractional mole weighting blending law. Referring to Figure 10, it was obvious that the first run was excluded, since it has a significantly higher initial propagation compared to the other two runs. The flame speed for the second and third run were measured to be 2.845 ms^{-1} and 2.700 ms^{-1}.

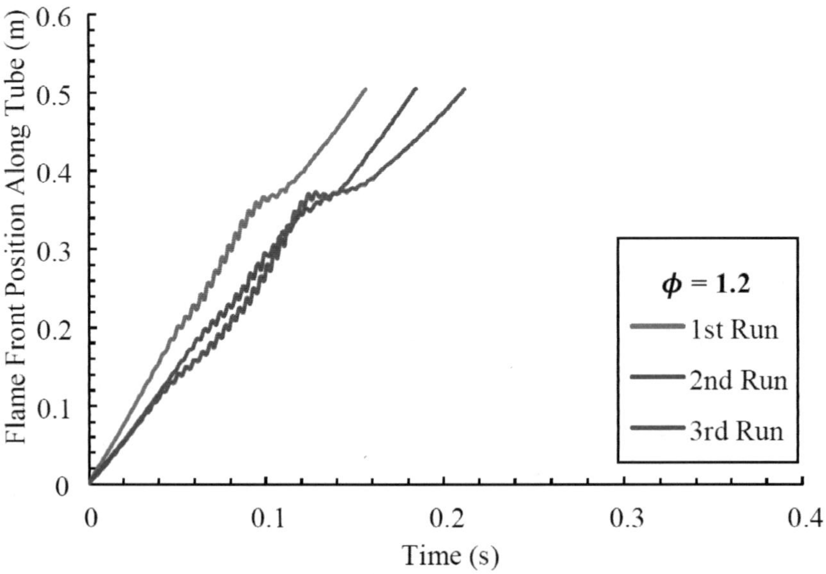

Figure 10. Tracked flame position for ϕ = 1.2 flames against time.

3.2 Measurement of laminar burning velocity

The measured flame speed from the previous section was used to calculate the laminar burning velocity. Initial flame front positions were compiled in Figure 11, which were used to calculate the flame speed in the previous section.

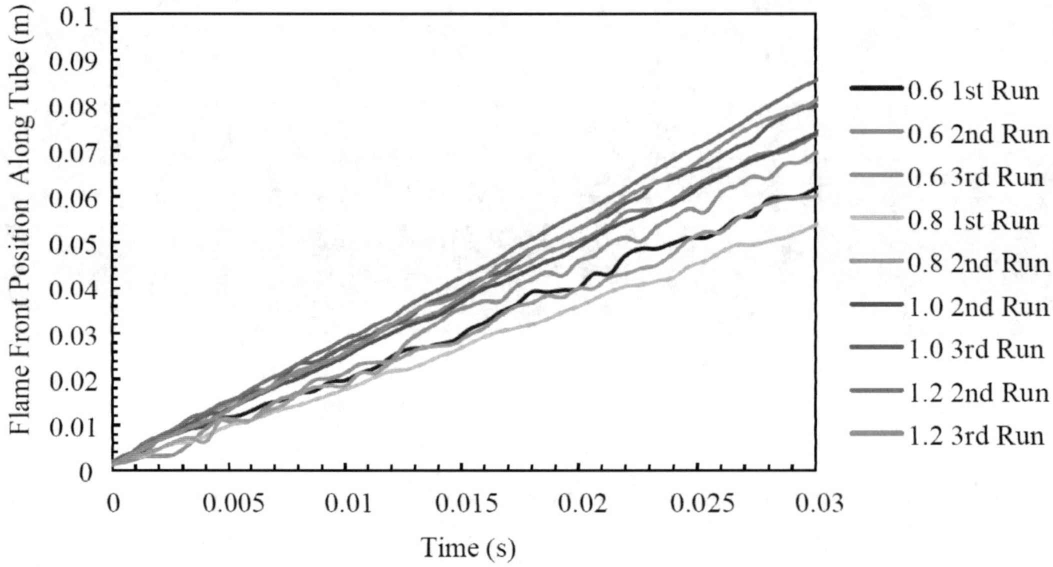

Figure 11. Initial flame front position for all flames, excluding anomalous runs.

The expansion ratio of each mixture was required in order to compute the laminar burning velocities, which was a common practice in previous research[18, 24]:

$$u_l = S_l \frac{\rho_b}{\rho_u} \tag{5}$$

where S_l = measured flame speed (m.s^{-1})
ρ_b = burned mixture density (kg.m^{-3})
ρ_u = unburned mixture density (kg.m^{-3})

Calculated values were then tabulated in table 3. Comparing them to the predicted laminar burning velocity of 0.6 ms^{-1}, the measured values were significantly lower. The average of all runs was calculated to be 0.368 ms^{-1} with a standard deviation of 0.056 ms^{-1}. It was observed that the ϕ = 0.6 flames had the closest value to the predicted value, which was 0.419 ms^{-1}, followed by ϕ = 1.2, ϕ = 1.0 and finally ϕ = 0.8.

Table 3. Fractional mole ratios calculated at different ϕ values.

ϕ	$\frac{\rho_b}{\rho_u}$	S_l (ms^{-1})			u_l (ms^{-1})			
		1	2	3	1	2	3	Avg.
0.6	0.183	2.055	2.516	2.318	0.375	0.459	0.423	0.419
0.8	0.154	1.791	2.002	N/A	0.276	0.308	N/A	0.292
1.0	0.138	N/A	2.463	2.661	N/A	0.339	0.366	0.352
1.2	0.138	N/A	2.845	2.700	N/A	0.394	0.374	0.384

The deviation observed in the experimental u_l values from the predicted values showed that other significant factors were in play. One of the main factors were thought to be the pressure pulsations which were inherent in flame tube rigs, caused by acoustic interactions within the tube, causing the oscillatory behavior in the middle of the tube. The initial propagation may appear to be steady, but low-pressure oscillations may have started forming, leading to the low experimental values.

Apart from low pressure oscillations, the flame tipping observed may have also affected the overall flame propagation. Flame tipping was thought to be caused by the density difference between the burned and unburned gasses, causing the flame to propagate in an asymmetric manner. Finally, the heat loss to the surrounding was thought to contribute to the slow flame speed, which is difficult to avoid with flame tube rigs.

4. Conclusion

In an attempt to produce similar flame propagation from mixtures with similar theoretical u_l of 0.6 ms^{-1} using the fifth fuel blending law proposed by Bradley *et al.* [9], it was discovered that the flames propagated with increasing speed from ϕ = 0.8 – 1.2 (0.292 ms^{-1}, 0.352 ms^{-1}, and 0.384 ms^{-1} respectively), with ϕ = 0.6 being an exception, which propagated the fastest among the mixtures. This was attributed to the high hydrogen content, $\bar{x}_{fh}$ = 0.899, which translates to 90% mass percentage hydrogen of the binary fuel. Apart from the speed difference, the flame oscillatory behaviour was also noticeably different, where most of them were oscillated significantly (ϕ = 0.8 – 1.2), while ϕ = 0.6 flames were only slightly vibrating. The blending law used may produce mixtures with similar theoretical laminar burning velocities, but it does not promise similar combustion behaviour, leading to different experimental laminar burning velocities.

References

[1] Çeper B A 2012 Use of Hydrogen-Methane Blends in Internal Combustion Engines. *Hydrogen Energy - Challenges and Perspectives* Minic D (ed) (London: InTechOpen)
[2] Ma F, Wang Y, Liu H, Li Y, Wang J and Zhao S 2007 Experimental study on thermal efficiency and emission characteristics of a lean burn hydrogen enriched natural gas engine *Int. J. Hydr. Ener.* **32** 5067–75
[3] Cho H M and He B Q 2007 Spark ignition natural gas engines-A review *Energy Conversion Management* **48** 608–18

[4] Pal A and Agarwal A K 2015 Comparative study of laser ignition and conventional electrical spark ignition systems in a hydrogen fuelled engine *Int. J. Hydr. Ener.* **40** 2386–95
[5] Van Lipzig J P J, Nilsson E J K, De Goey L P H nd Konnov A A 2011 Laminar burning velocities of n-heptane, iso-octane, ethanol and their binary and tertiary mixtures *Fuel* **90** 2773–81
[6] Payman W and Wheeler R V 1922 The composition of gaseous fuels in relation to their utilisation *Fuel in Science and Practice* **1** 185–96
[7] Di Sarli V and Di Benedetto A 2007 Laminar burning velocity of hydrogen-methane/air premixed flames *Int. J. of Hydr. Ener.* **32** 637–46.
[8] Bradley D, Habik S E D and El-Sherif S A 1991 A generalization of laminar burning velocities and volumetric heat release rates *Combustion and Flame* **87** 336–45
[9] Bradley D, Lawes M and Mumby R 2017 Burning velocity and Markstein length blending laws for methane/air and hydrogen/air blends *Fuel* **187** 268–75
[10] Hirasawa T, Sung C J, Joshi A, Yang Z, Wang H and Law CK 2002 Determination of laminar flame speeds using digital particle image velocimetry: Binary fuel blends of ethylene, n-butane, and toluene *Proceedings of the Combustion Institute* **29** 1427–34
[11] Spalding D B 1957 I. Predicting the laminar flame speed in gases with temperature-explicit reaction rates *Combustion and Flame* **1** 287–295.
[12] Spalding D B 1957 II. One-dimensional laminar flame theory for temperature-explicit reaction rates *Combustion and Flame* **1** 296–307
[13] Morley C 2005 *GASEQ* Version 0.79 (www.gaseq.co.uk: Morley C)
[14] Pennington P *Flames in Tubes[master's thesis]* 2009 (Sheffield: University of Sheffield)
[15] Amaludin N A 2020 *Flame Instabilities in Open Ended Tubes [dissertation]* (Sheffield: University of Sheffield)
[16] Vision Research 2016 *Phantom Camera Control* Version 2.7 (New Jersey: Vision Research)
[17] Ebieto C E 2017 *Dynamics of Premixed Flames in Tube[dissertation]* (Sheffield: University of Sheffield)
[18] Gu X J, Haq MZ, Lawes M and Woolley R 2000 Laminar burning velocity and Markstein lengths of methane–air mixtures *Combustion and Flame* **121** 41–58
[19] Glassman I and Yetter R A 2008 *Combustion* 4th Edition (Cambridge: Academic Press)
[20] Yang J, Mossa F M S, Huang H W, Wang Q, Woolley R and Zhang Y 2015 Oscillating flames in open tubes *Proceedings of the Combustion Institute* (The Combustion Institute) 2075–2082.
[21] Guénoche H Flame Propagation in Tubes and in Closed Vessels in: *Nonsteady Flame Propagation* 107–181
[22] Ebieto C E, Amaludin N A and Woolley R 2015 Methane / Hydrogen / Air Flame Oscillations in Open Ended Tubes *Proceedings of the 25th International Colloquium on the Dynamics of Explosion and Reactive Systems* (The Institute for the Dynamics of Explosions and Reactive Systems)
[23] Markstein G H 1964 *Nonsteady Flame Propagation* (New York: Pergamon Press)
[24] Hu E, Huang Z, He J and Jin C, Zheng J 2009 Experimental and numerical study on laminar burning characteristics of premixed methane-hydrogen-air flames.*Int. J. of Hydrogen Energy* **34** 4876–88

Vortex profile analysis under different diffuser size for inlet channel of gravitational water vortex power plant

F M Tamiri[1*], E C T Yeo[1], M A Ismail[2*]

[1]Faculty of Engineering, Universiti Malaysia Sabah, Jalan UMS, 88400 Kota Kinabalu Sabah, Malaysia
[2]Centre of Research in Energy and Advanced Materials, Universiti Malaysia Sabah, Jalan UMS, 88400 Kota Kinabalu, Sabah, Malaysia

*Corresponding Author: lanz_mr@ums.edu.my, fadzlita@ums.edu.my

Abstract. Hydropower is a renewable technology to store the amount of electricity which is the least expensive. Gravitational Water Vortex Power Plant is an ultra-low head micro hydropower system working ranging from 0.7 m to 2m without having the needs of a large reservoir and installation area. Several researches have been conducted on its basin configuration, orifice diameter, blade configuration, the geometry of the basin shape but not onto the addition of the diffuser at the inlet channel. The function of the diffuser is to direct the water into the basin allowing the water vortex to travel towards the tangential direction where this phenomenon will increase the rate of speed flow through the turbine. The simulation results showed that the addition of the diffuser has significantly increased the tangential velocity and the kinetic energy of the vortices. The increase in the velocity of the flow increased the height of the vortex which also led to the increase in the strength of the vortex and affects the vortex uniformity.

Keywords: microhydro, diffuser size, flow velocity, strength of the vortex, vortex uniformity

1. Introduction

Electricity demand in the ASEAN region has been increasing by 70% compared to the energy demand in the year 2000 [1]. ASEAN's demand for electricity is growing rapidly fast due to the outstandingly high economic growth prospect of the region [1, 2]. At the same time, carbon dioxide emits more than tripled in line with the growing share of fossil fuels in the energy which brought concerns regarding climate change in Southeast Asia. While awareness is needed for action on climate change, the high value of oil and gas, and the outstanding potential of renewable-based electricity have brought about potent causes for expanding renewable electricity generation. Renewable based generated electricity offers numerous energy security and environmental benefits, including greater variety in the power mix and reduced local air pollution. Hydro assets, especially the Philippines, Thailand, and Vietnam, have the greatest potential. With more than a fifth of the capacity realized, hydro has been the most exploited of the region's renewable energy resources [3].

1.1 Gravitational water vortex power plant (GWVPP)

Gravitational Water Vortex Power Plant is considered as a micro hydro power plant as it has an optimum power generation less than 100 kW and characterized as an ultra-low head micro hydropower system

Content from this work may be used under the terms of the Creative Commons Attribution 3.0 licence. Any further distribution of this work must maintain attribution to the author(s) and the title of the work, journal citation and DOI.
Published under licence by IOP Publishing Ltd

IOP Conf. Series: Materials Science and Engineering 1217 (2022) 012014 doi:10.1088/1757-899X/1217/1/012014

working in the range of 0.7 m to 2 m without having the needs of a large reservoir and installation area. GWVPP is a kind of green technology which the hydropower system generates electricity by using the natural characteristic of the water to form a free vortex surface. Water vortex will create a turbine's swirl flow [4]. The advantages of GWVPP are that this type of power plant is environmentally friendly and the generation of electricity is from ultra-low hydraulic pressure. Furthermore, this form of power plant can be built on a river or steam to produce electricity for a few houses targeted those who stayed in the rural. The cost of developing and generating electricity in GWVPP is very low as compare to others hydro power technologies [5, 6]. Past researches paid more attentions on optimizing the turbine inlets, outlets, design and the geometry of the basin.

Gravitational Water Vortex is considered as a useful application of the Free Surface Vortex (FSV), which is a significant phenomenon in the field of hydraulic engineering. It can either be considered as a destructive or beneficial source according to its locality. Gravitational water vortex power plants consist of a canal, a basin structure, and a turbine. Water through the canal is tangentially fed into the round basin structure to form a powerful water vortex. The kinetic energy of the water vortex is then converted to electric energy through the turbine runner located at the centre of the water vortex. An orifice is made at the bottommost of the basin through which the water vortex discharged [7]. The maximum efficiency obtained by researchers and commercial companies is still very low, about 30% compared to other conventional low head turbines. This new technology has great potential to be developed and to be optimized in both performances of hydraulic and mechanical efficiency to sustain relatively high efficiency at low heads and with small to medium flow rates. There is a lot of investigation carried out through experimental work and simulation analysis for the basin configuration, orifice diameter, blade configuration, geometry of the basin shape and the runner of the gravitational water vortex power plant but not onto the addition of the diffuser at the inlet channel. The function of the diffuser is to direct the water into the basin allowing the water vortex to travel towards the tangential direction where this phenomenon will increase the rate of speed flow through the turbine.

1.2 Diffuser in GWVPP

A uses of diffuser lead to a continuity equation which state that there is a steady flow in a channel and the principle of conservation of mass applies to the system. The mean of velocities at all cross sections with equal areas are equal and if the areas are not equal, the velocities are inversely proportional to the areas of the respective cross sections. Thus, assuming the flow of water is constants in a channel array, the product of the area and the velocity will be the same for every two cross sections within that range. The equation is given by $Q = A_1V_1 = A_2V_2$, where Q is the volumetric flow rate, A is the cross-sectional area of flow and V is the mean of velocity [8]. The functions of diffuser acts as an output power accelerator for horizontal axis water current turbine for the low flow velocity. The usage of diffuser configuration is expected to increase the rate of speed flow through the turbine.

According to Dhakal *et al.* studies [9], notch inlet width, cone angle and channel height have the most significant effects on the vortex velocity. They suggested that smaller value of notch inlet width of the diffuser and higher value of the cone angle and the channel height increased the vortex performance. Longer value of notch length or length of inlet channel is also recommended to prevent unwanted loses.

2. Research methodology

2.1 Experimental setup

Figure 1 below is the experimental setup, where the design of the water vortex generation system can be divided into several parts such as centrifugal pump, flow rate indicator, turbine, diffuser, basin and also storage tank. The material used for the diffuser, basin and the inlet channel is stainless steel. There are two centrifugal pumps of 1.5 kW that were installed to this experiment setup, but one centrifugal pump is sufficient to supply enough power to run. A water flow rate indicator was installed at the pipe to measure the water flow rate which the flow rate of water can be controlled using valve. Four sets of diffusers with different size were tested and to be installed at the channel inlet before conducting the experiment, as shown in figure 2. The dimension for each diffuser is shown in table 1.

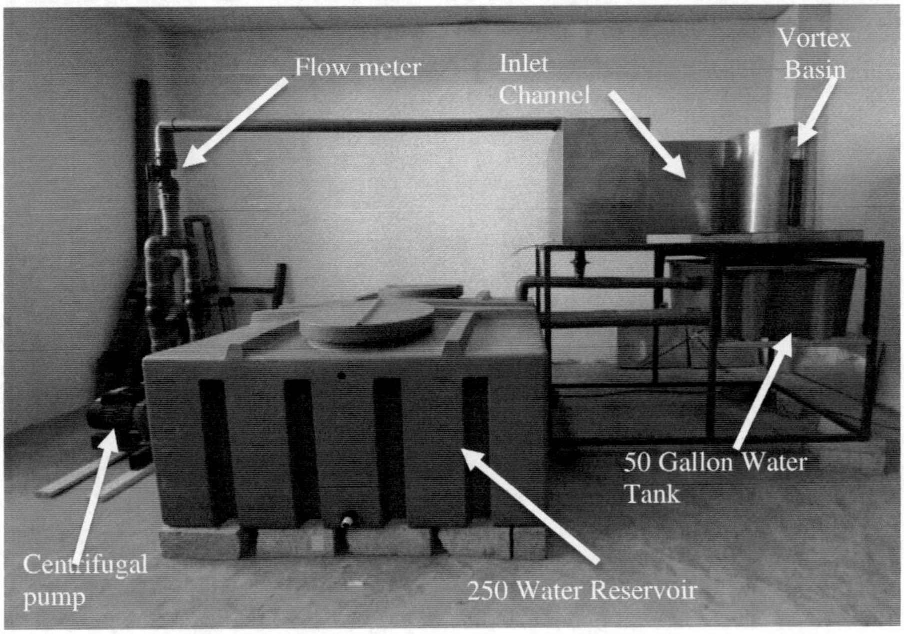

Figure 1. Full set up of vortex turbine generator.

Figure 2. Diffuser configuration.

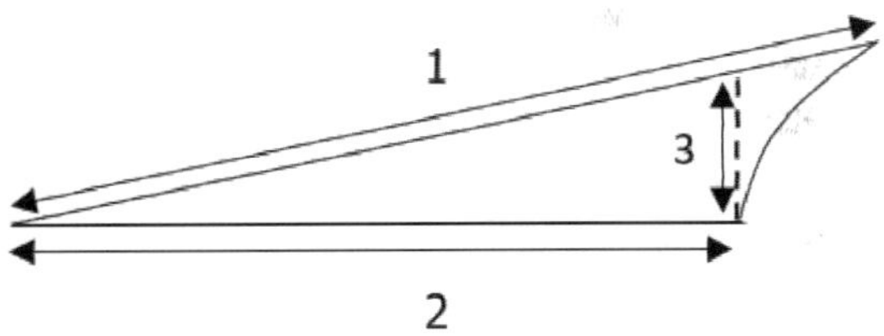

Figure 3. Design of the diffuser configuration.

Table 1. The dimension of the diffuser configuration.

Configuration of Diffuser	Length (cm) 1	2	3	Angle (°)
A	58.5	51.0	9.0	10.00
B	55.9	51.0	7.0	7.82
C	53.7	51.0	5.0	5.60
D	52.3	51.0	3.0	3.37

IOP Conf. Series: Materials Science and Engineering 1217 (2022) 012014 doi:10.1088/1757-899X/1217/1/012014

For data analysis in experimental work, the height of the water vortex will be measured from the bottom of the basin, and the ruler is placed at the wall of the basin. The derivation of an equation by Odgaard [10] presented the relationship between the height of the vortex and the maximum tangential velocity, as shown below.

$$H = 1.7\,\frac{V_m^2}{g} \tag{1}$$

Where, H is the vortex height
V_m is the maximum tangential velocity
g is the gravitational acceleration

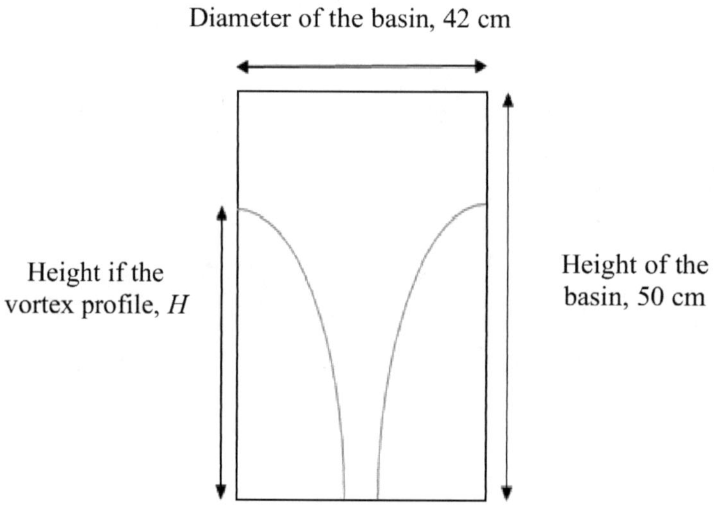

Figure 4. The dimension of the basin.

2.2 ANSYS CFX

2.2.1 Meshing. Geometry is an essential factor when it comes to the optimization of shape where accurate geometry representation and the use of adaptive analytical techniques are essential for computationally efficient Finite Element simulations [11]. The tetrahedral mesh was selected in all the flow regions as it was the best for solving complex geometry topology in order to capture the flow physics inside the domains. Hence, the number of elements for the volute elements was found to be 1105576 and the number of nodes was 196825. The best meshing generation size would be 0.012 m for the simulation. One of the key challenges in assessing grid sensitivity with respect to the numerical simulation results is to find the right balance between mesh resolution and calculation time. Therefore, an independent mesh analysis was performed at an earlier stage of simulation analysis. It is essential to select a suitable mesh number to acquire a set of accurate models and results. Mesh independent analysis was performed to evaluate the rate of convergence, solution accuracy and computational time. Maximum Global Velocity, m/s is an important and useful parameter that can be observed at every step of mesh refinement.

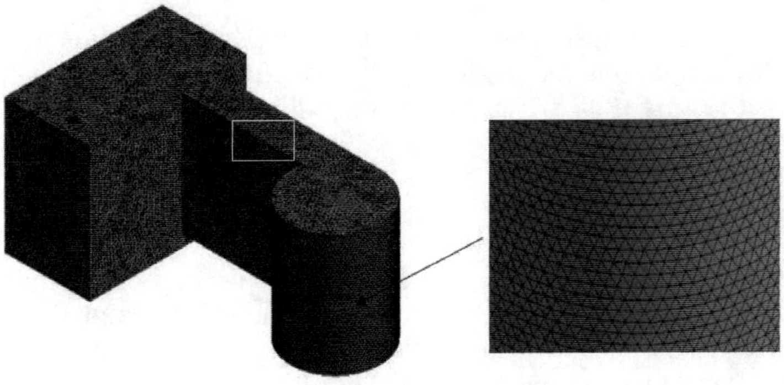

Figure 5. Meshing topography

2.2.2 Solver and boundary condition. In this simulation, two types of fluid were used, which are 'Air' and 'Water' in this setting. The fluid models were set a homogeneous model where the free surface model is set to standard. The Surface Tension Coefficient was set at 0.072 Nm^{-1} and the primary fluid used in this simulation is water.

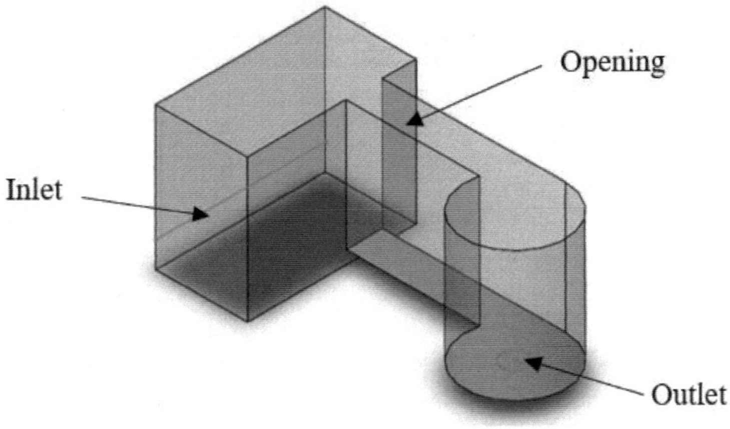

Figure 6. Boundary Condition for GWVPP.

As shown in figure 6, the boundary for inlet was set at the bulk mass flow rate of 4 kg/s where the air has 0 volume fraction and 1 water volume fraction. For the opening (free surface), the relative pressure was set at 0 Pa where the air has 1 volume of fraction and 0 water volume fraction. The boundary condition for the outlet (free surface) was set at the relative pressure of 0 Pa where the air volume fraction is 0 and the water volume fraction is 1.

2.2.3 Convergence criteria. In the studies, the fluid dynamics problems were presented in non-linear equations governed by the Navier-Stroke equations. They were solved iteratively by numerical methods which lead to convergence criteria. The residuals for each computational domain were plotted at the end of the time step to measure the convergence solution. In order to have the following convergence criteria, the Residual Root Mean Square (RMS) Error-values must be reduced to 1×10^{-4}, the domain had imbalances of less than 1% of its value or the values of interest had reached a steady solution. Thus, the meshing of this simulation was set at the maximum iteration of 500. Therefore, the variable values of Roots Means Square per volume was stable after the 400th iteration. Thus, the most suitable iteration used in this project was the 500th iteration for each simulation.

IOP Conf. Series: Materials Science and Engineering 1217 (2022) 012014 doi:10.1088/1757-899X/1217/1/012014

2.3 Validation

To validate the results attained from simulation, the prediction of the simulation was compared with the experimental data. The comparison was performed by measuring the height of the vortex from the bottom of the basin, at the wall of the basin. Validation will be accepted if the CFD simulation shows that the simulation matches the height of the vortex profile carried out on experimental work.

3. Results and discussions

3.1 Validation of height of the vortex

Figure 7 indicates the 3D streamline of the water flow in the setup. Validation was done from the experimental results to match the height of the simulation analysis. The convincing result of the validation will lead the project in the right direction.

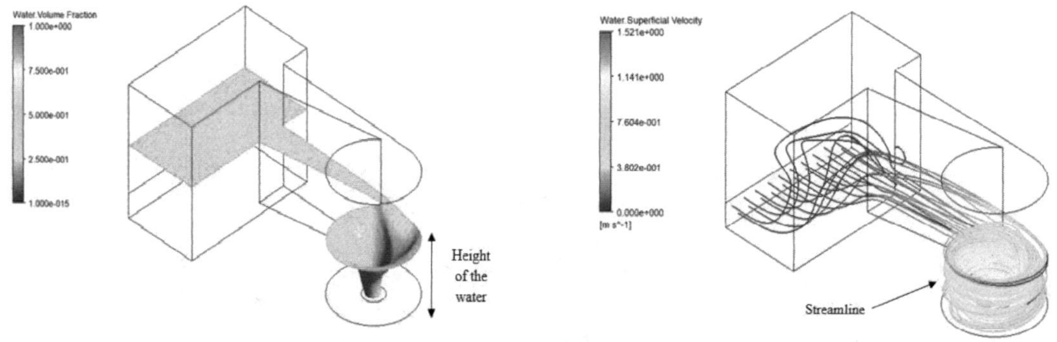

(a) (b)

Figure 7. (a) Isosurface (b) Streamline flow of water in Gravitational Water Vortex Power Plant.

Isosurface set at the value 0.5 indicated the height of the water. To validate the data, the experimental data were used to validate simulation data. However, it was found that there is difference between simulation and experimental data, recorded at 8%. The difference may be contributed by many factors, however one of the most prominent observation was due to the fluctuating of water height which leads to the difficulty in taking exact height at steady state condition. As the water were flowed irregularly, the only method to take a higher accuracy was to take several readings and get the average height of the water.

3.2 Comparing water superficial velocity at different depth for plane X and plane Z

The fluid flow layers were observed in Plane X as the fluid flow layer was spiral in a vortex and had tangential velocities greater than the radial and axial velocities when the turbine blade was placed at a position of 35% above from the bottom of the basin [12]. This arrangement allowed the water to transfer full kinetic energy to the blade in generating electricity. In order to determine the most optimum performance of diffuser, data of water superficial velocity was studied. The investigation was carried out with different diffuser configurations at a different depth ranging from 0.06 m to 0.22 m.

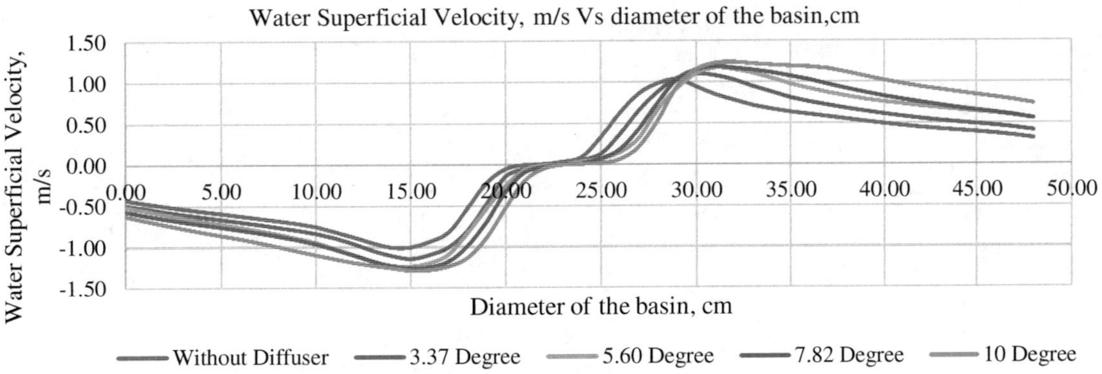

Figure 8. Water Superficial Velocity at the depth of 0.06 m for Plane X.

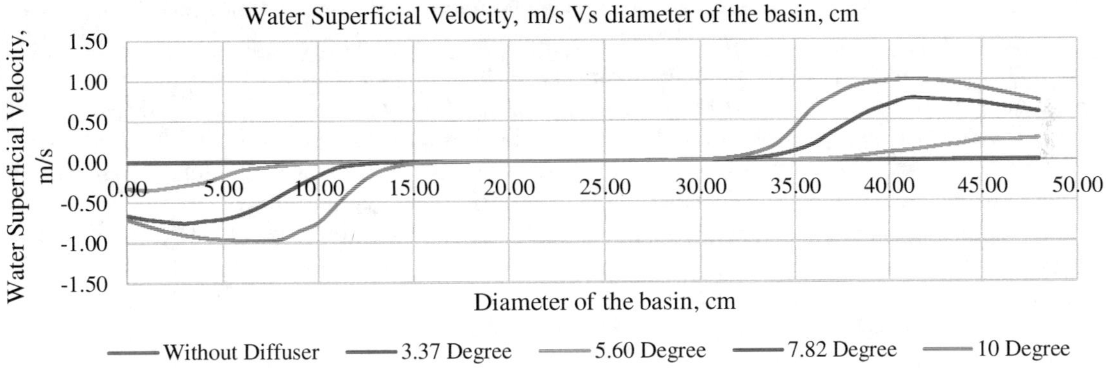

Figure 9. Water Superficial Velocity at the depth of 0.22 m for Plane X.

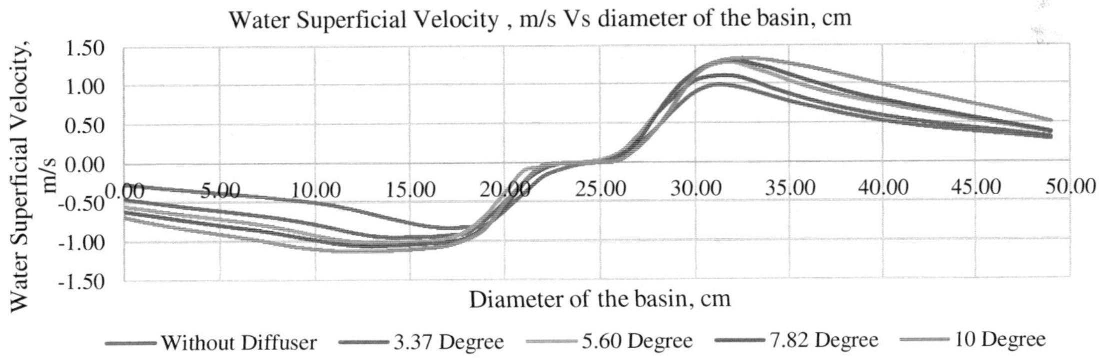

Figure 10. Water Superficial Velocity at the depth of 0.06 m for Plane Z.

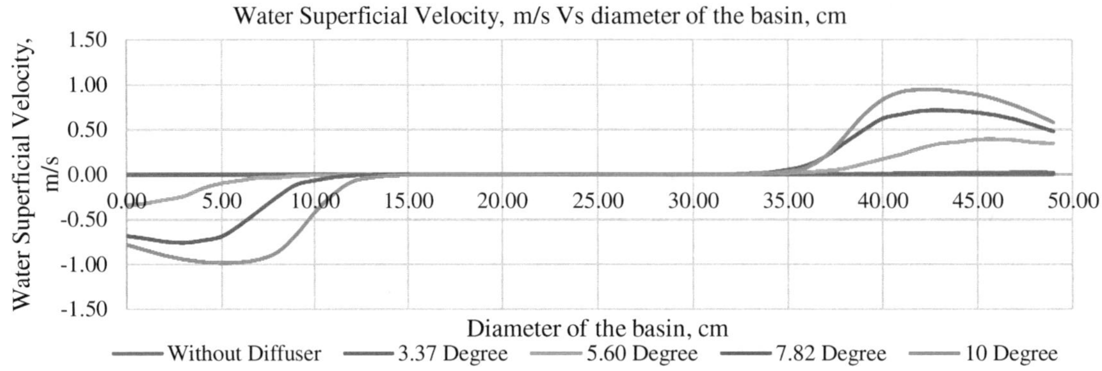

Figure 11. Water Superficial Velocity at the depth of 0.22 m for Plane Z.

Based on figure 8, the diffuser A (10°) had the maximum velocities at 0.06 m, 0.10 m, 0.14 m, 0.18 m and 0.22 m followed by diffuser B (7.82°), diffuser C (5.60°), diffuser D (3.37°) and with no diffuser. In addition, the water velocity with no diffuser added at the inlet channel decreases tremendously as the depth reached 0.18 m above the basin. Moreover, figure 9 illustrated that the water superficial velocity is zero at a depth of 0.22 m above the basin for diffuser D and without diffuser. As for diffuser C, the water velocity decreases. The water velocity for diffuser A remains at high velocity. Thus, the observation goes the same to Plane Z as shown in figure 10 and figure 11.

As shown in figure 9 and figure 11, there was not a need to study the depth above 0.22 m as the velocity showed a drastic decrease in Diffuser B (7.82°). Thus, according to Dhakal *et al.* [12] investigation, which the turbine was the most optimum when placed 35% above the basin. Diffuser A shown to be the most suitable to be placed at the inlet channel as the height of the water can reach up to 0.28 m provided that the velocity of the water was still solid enough to provide kinetic energy to the turbine.

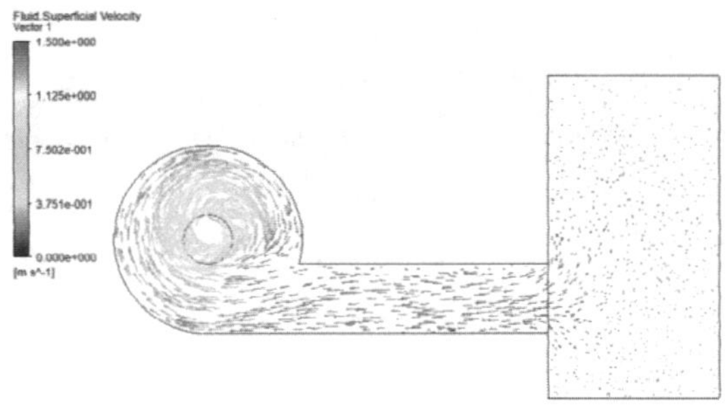

Figure 12. Velocity vector of the flow without diffuser.

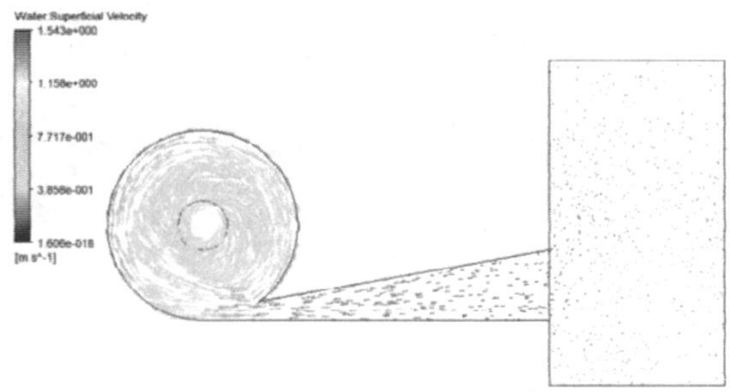

Figure 13. Velocity vector of the flow with diffuser.

Based on figure 12, the velocity vector showed that the vortices were skewed towards the wall of the basin. After the insertion of the diffuser at the inlet channel based on figure 13, it can be seen that the vortices were more uniform with greater vortices. The diffuser caused an increase in the water superficial velocity. Vorticity leads to the increase of fluid velocity which decreased the viscosity, which disintegrates the fluid parcel moving along the streamline. This action has let the vortices generate, where vorticity, being twice the angular velocity is a measure of local spin in the fluid element given by the curl of the velocity. This phenomenon has resulted in an increase of vortex height due to the increase in the tangential velocity of the fluid corresponding in the decrease in radial and axial velocities.

3.3 Comparing the flow pattern for different diffuser configuration

In the following section, the vortex profile was analysed, resulting in different diffuser configurations. In this following segment shows Plane X (1.05 m) and Plane Z (-0.33 m) where 5 lines were drawn at different depth. The red area shows the vortex flows outwards and the blue area flows inward through a plane. As shown in figure 14, the deeper colour of red and blue shows the fastest velocity flow of the vortex and the data was recorded as shown in figure 14. Based on the figures, the water vortex swirls around an empty core in decreasing radius like a 'V'-shape. A water vortex can be characterized as a flow pattern with concentric circle streamlines. Since all the fluids have viscosity, there is no flow that is really irrational. However, a diffuser may increase in the velocity of the flows which helps to overcome the viscous effects and later be assumed that they are invisible and irrotational due to the negligible viscous effects Thus, the fluid particles do not rotate in a free irrotational vortex, as they translate into circular paths around the vortex centre.

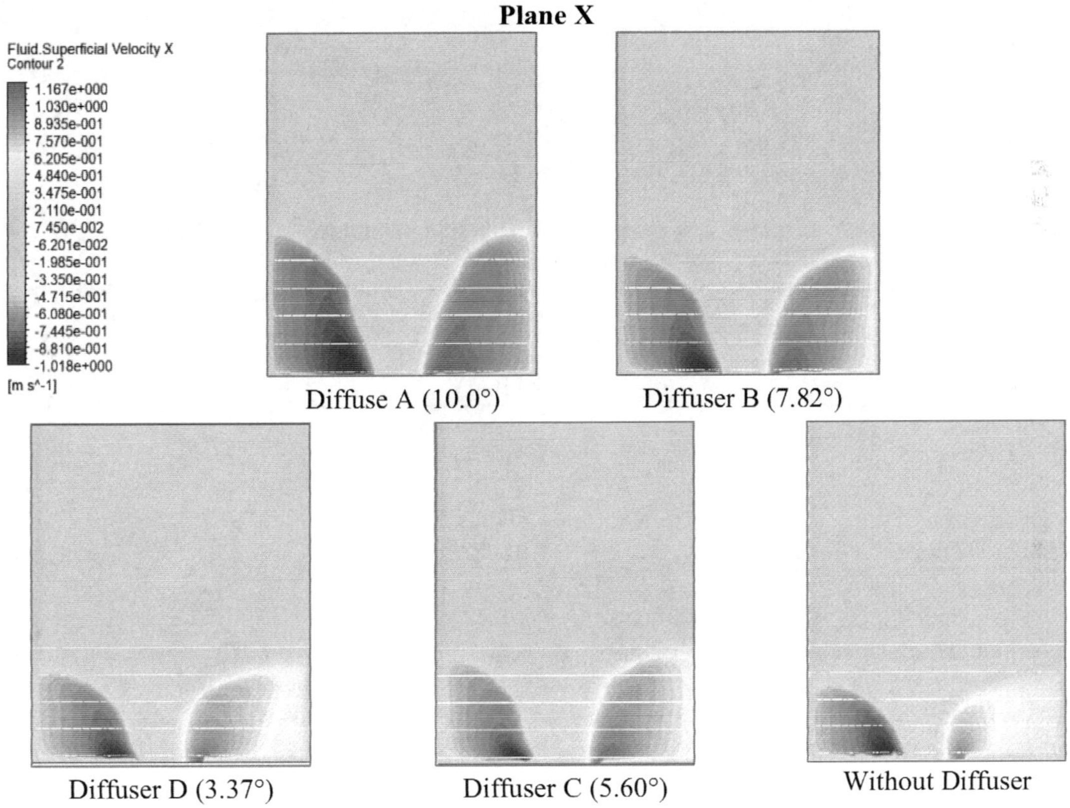

Figure 14. Water vortex profile for Plane X.

The placement of the diffuser distorted the vortex profile, as shown in figure 14. It shows an increase in the height of the vortex and the strength of the vortex. When the diffuser angle increases, there is an increase in the vortex profile. Different diffuser sizes lead to different velocity values. The pattern was formed when water was flowing through the inlet channel and accelerate through the diffuser and spin around the basin which formed the vortex. The increase in the vortex height was caused by the increasing angle of the diffuser which reduce the area of the inlet channel, which directed and accelerated the water towards the tangential side of the vortex without disrupting the air core.

IOP Conf. Series: Materials Science and Engineering 1217 (2022) 012014 doi:10.1088/1757-899X/1217/1/012014

3.4 Comparing the maximum velocity of the circulation of water in a basin with different diffuser configuration

Based on figure 15, Plane X and Plane Z are used to study the maximum flow of the water circulating around the basin with different diffuser configuration.

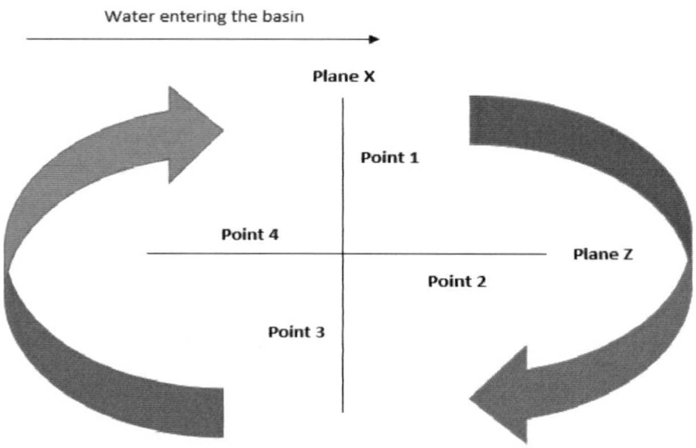

Figure 15. Circulation of the water.

Table 2. Maximum Velocity at each point.

Diffuser Configuration	Maximum Outward (Point 1) Velocity in Plane X, m/s	Maximum Outward (Point 2) Velocity in Plane Z, m/s	Maximum Inward (Point 3) Velocity in Plane X, m/s	Maximum Inward (Point 4) Velocity in Plane Z, m/s
Diffuser A	1.25773	1.36415	1.30723	1.15716
Diffuser B	1.21878	1.31878	1.27525	1.08014
Diffuser C	1.25113	1.3233	1.29718	1.05247
Diffuser D	1.19647	1.19663	1.15759	0.970577
Without Diffuser	1.21205	1.02783	1.01755	0.880096

The entry velocity to the basin without a diffuser is 1.21205 m/s. When the diffuser is placed on the inlet channel, the entry velocity to the basin increases as shown in table 2. As the size of the diffuser increases, the flow of the water into the basin increases. The studies had shown that the maximum velocity in plane X is 1.21205 m/s and the velocity of the circulation decreases when there is no diffuser configuration. The second set of studies shows the maximum velocity for diffuser configuration D is 1.19647 m/s at point 1 and the velocity is maintained at around 1.19663 m/s at point 2 and decreases as it reached point 3.

In the third set of studies shows the maximum velocity for diffuser configuration C is 1.25113 m/s in point 1 and increases up to 1.3223 m/s at point 2 and decrease slightly at 1.29718 m/s at point 3 and to 1.05247 m/s at point 4. In the fourth set of studies for diffuser B, the maximum velocity began at a low velocity which is 1.21878 m/s but increases steeply to 1.31878 m/s at point 2 and decreases to 1.27525 m/s as it reached point 3 and 1.081014 m/s at point 4.

In the fifth set of studies for diffuser A, it was observed that the maximum velocity at all 4 points is the highest among others set of studies. It began at the velocity of 1.25773 m/s after the inlet channel and reached a maximum velocity of 1.36415 m/s at point 2 and decrease slightly at point 3 at 1.30723 m/s and to 1.15716 m/s at point 4.

Thus, according to the simulation, it can explain that the crowding-out effect that results in the increase of the height of the water at higher velocity. The water is accelerating at a higher velocity when the angle of the diffuser is higher. The water is accelerating after the water enters the inlet channel and accelerates the water to point 2, which also provides higher kinetic energy to water to circulate around basin which can lead to an increase in the efficiency of the rotating turbine. Last but not least, diffuser configuration A gives the most optimized performance to this experiment.

The development of a vortex occurs when the transition from the open channel to pressure flow is not smooth and uniform. When the flow does not maintain a gradual transition that attempts to keep a uniform velocity distribution and acceleration, the vortices will occur. Vortex formation is caused by the non- uniform approach flow to the intake, shear layers of high-velocity gradients and the rotation caused by an obstruction in the approach of the flow. As for the uniformity of vortex flow in the set up without diffuser, the uniformity of the vortex is not uniform due to the low kinetic energy flowing into the entry of the flow and it is affected by the shear layers. Thus, with the help of the diffuser A, it can be observed that uniformity of the vortex flow remains its strength and only change slightly in the shape. Thus, it provides a greater strength of vortex.

4. Conclusion

Based on the findings on the research, it is concluded that inlet channel with diffuser with higher degree of angle will maximize the flow of the water through the inlet channel. The diffuser act as an accelerator which provide higher kinetic energy to the water to circulate around basin which can leads to an increase in the efficiency of the rotating turbine. The placement of the diffuser also distorted the vortex profile which developed a fully developed air-core through the vortex height. The following phenomenon showed an increase in the height of the vortex and the strength of the vortex. When the diffuser angle increased, there is an increase in the vortex profile. The increase in the height of the vortex will produce a greater vortex strength. As the vortex strength increased, the motion of the water vortex tends to travel in a tangential direction instead of an axial direction. The increase in the vortex height was caused by the increasing angle of the diffuser which reduce the area of the inlet channel, which directed and accelerated the water towards the tangential side of the vortex without disrupting the air core.

Since the most optimum position of the blade was when placed 35% above the basin. Diffuser A with the angle of 10° has shown to be the best diffuser to be placed at the inlet channel as the water can reach up to 0.28m which the velocity of the water was still solid enough to provide kinetic energy to the turbine. It was observed that it has the highest maximum velocity at all 4 points compared to others set of studies. It began at the velocity of 1.25773 m/s after the inlet channel and reached a maximum velocity of 1.36415 m/s at point 2 and decrease slightly at point 3 at 1.30723 m/s and to 1.15716 m/s at point 4.

Acknowledgement

The authors would like to thank the Ministry of Higher Education of Malaysia for funding this research under the Fundamental Research Grant Scheme: FRGS/1/2020/STG07/UMS/02/2 and Universiti Malaysia Sabah through their financial assistance and facilities support.

References

[1] Liu Y and Noor R 2020 Energy Efficiency in ASEAN: Trends and Financing Schemes *ADBI Working Paper 1196* Tokyo: Asian Development Bank Institute. Available: *https://www.adb.org/publications/energy-efficiency-asean-trends-financing-schemes*.
[2] Chang Y and Li Y 2013 Power generation and cross-border grid planning for the integrated ASEAN electricity market: A dynamic linear programming model *Energy Strategy Reviews* **2**(2) 153-160
[3] Shadman F 2016 Drought and energy security in key ASEAN countries *Renewable and Sustainable Energy Reviews* **53** 50-58
[4] Power C, McNabola A and Coughlan P 2016 *J. of Clean Ener. Tech.* **4** 112-119
[5] Zotlöeterer F 2014 Smart Energy Systems. Available: *http://www.zotloeterer.com/welcome/gravitation-water-vortex-power-plants/zotloeterer-turbine/*

[6] Rahman M M, Tan J H, Fadzlita M T and Wan Khairul Muzammil A R 2017 A Review on the Development of Gravitational Water Vortex Power Plant as Alternative Renewable Energy Resources *IOP Conf. Series: Mat. Sci. and Eng.* 217, 012007.

[7] Shabara O B Y and Ahmed Y M 2015 CFD Simulation of Water Gravitation Vortex Pool Flow for Mini Hydropower Plants *Jurnal Teknologi* **74**(5).

[8] Dooge J C I and Napiorkowski J J 1987 The Effect of the Downstream Boundary Conditions in the Linearized St Venant Equation *The Quarterly J. of Mech. and Appl. Math.* **40**(2) 245-256

[9] Dhakal S, Timilsina A B, Dhakal R, Fuyal D, Bajracharya T R and Pandit H P 2014 *Int. Conf. on Tech. and Innov. Management & IOE Graduate Conf.* Kathmandu Nepal pp 380-386

[10] Odgaard A J 1986 Free-surface air core vortex *J. of Hydraulic Eng.* **112**(7) 610-20

[11] Marco O, Ròdenas J J, Navarro-Jimènez J M and Tur M 2017 Robust h-adaptive meshing strategy considering exact arbitrary CAD geometries in a Cartesian grid framework *Computers & Structures* **193** 87-109

[12] Dhakal S, Nakarmi S, Pun P, Thapa A B and Bajracharya T R 2014 Development and Testing of Runner and Conical Basin for Gravitational Water Vortex Power Plant *J of the Inst. of Eng.* **10**(1) 140–148

Simulation of solar PV surface temperature with dimpled fin cooling

M F Ibrahim[1], M S Misaran[1*], N A Amaludin[2]

[1]Faculty of Engineering, Universiti Malaysia Sabah, Jalan UMS, 88400 Kota Kinabalu, Sabah, Malaysia
[2]Centre of Research in Energy and Advanced Materials, Faculty of Engineering, Universiti Malaysia Sabah, Jalan UMS, 88400 Kota Kinabalu, Sabah, Malaysia.

[*]Corresponding authors: suffian@ums.edu.my

Abstract. A significant number of cooling technologies have been developed to maintain the PV module temperature within subscribed limits. This paper represents the simulation study of active cooling forced air convection with fins attached to the back of the solar panel using CFD SimScale software. It has been first carefully validated against experimental and numerical results available in the literature. The number of fins and the shape of perforated and dimpled in each fin were varied to compare cooling performance. Three types of fins were adapted into this simulation: traditional fins, circular, and triangle perforated/dimpled fins. The effect of solar irradiation and velocity inlet was also reviewed by applying the nominal operating condition from the experimental works. Results indicated that fin channels are a very effective cooling technique, which significantly reduces the average temperature of the PV cell, especially when increasing the number of fins from 20 to 26 fins. Also, the results show that the dimpled triangle fin had the highest average temperature drop with a percentage difference of 6% compared with the solar panel cooling with traditional fins.

1. Introduction

Environmental factors such as wind velocity, ambient temperature, humidity, dust, and solar radiation are the most frequent factors that affect the efficiency of a solar panel [1]. However, the influence of the panel's temperature on solar panel efficiency remains one of the significant drawbacks of this technology. As such, the cooling of photovoltaic panels can be improved if the PV is equipped with metallic materials with fins back surface panels to ensure a much higher circulation of air [2].

In literature, modifications to optimize the performance of solar PV with cooling fin were proposed to enhance the flow and consequently increase efficiency. Saadi *et al*. [3] studied the effective air forced cooling of photovoltaic solar cells by inserting triangular ribs at the back of the solar cell. In this study, it can be concluded that the higher the number of fins, the higher the rate of heat transfer of the solar cell. Jaemin Kim & Yujin Nam [4] analysed a study on the cooling effect of attached fins on PV using CFD simulation. The results of the CFD simulation show that the installation of the fins influenced the airflow and temperature distribution at the bottom of the PV module, where the temperature difference was reduced to less than 4.6 °C when the fins were mounted. This means that the area exposed to air has a more significant impact on cooling efficiency.

The optimization process of solar panel forced air cooling with fin has reached the stage where the fin geometrical configuration was modified to enhance the heat transfer rate for better mixing of air

Content from this work may be used under the terms of the Creative Commons Attribution 3.0 licence. Any further distribution of this work must maintain attribution to the author(s) and the title of the work, journal citation and DOI.
Published under licence by IOP Publishing Ltd

which directly increases the temperature drop of the solar PV panel. Arifin *et al.* [5] and Popovici *et al.* [6] proposed a CFD simulation analysis on the cooling process using a perforated cooling fin. As a result, the average temperature of PV panels with perforated cooling fin heat sinks was lower than in PV panels without perforated cooling fin heat sinks. Thus, the temperature contour of the PV panel with and without aluminium heat sinks with perforated fins with a heat flux of 1000 W/m^2 indicates a substantial drop in temperature down to 12.5°C, and an improvement in the heat transfer output from the panel to the air when aluminium heat sinks with perforated fins are used.

Gupta *et al.* [7] conducted a simulation study comparing the heat augmentation from a plane, dimpled and perforated dimple rectangular fins using Ansys CFD software. The fins' length, height, and thickness were fixed at 100 mm, 120 mm, and 3 mm, respectively. In this analysis, for the same base temperature, the perforated dimpled fins show a higher temperature drop than the dimpled fin and the plane fin, particularly for the triangle shape of the perforated dimpled fin. However, a dimpled fin was never used to cool down solar PV surface temperature; thus, the objective of this paper is to investigate the temperature drop of solar PV surface with dimpled cooling fin and compares the result with perforated fins using the simulation method.

2. Methodology

2.1 3D model

This study designed the perforated and dimpled cooling fin and simulated the heat transfer using CFD software. The solar panel design was drawn based on the actual size solar panel (845 mm x 515 mm x 35 mm), and the fins were varied at; circular and triangular shapes and perforated and dimpled designs as shown in figure 1. Suitable dimensions of cooling fins for heat transfer have been discussed by Raina *et al.* [8] and is adapted in this study as shown in table 1. The aluminium fin was mounted on the back of the solar panel, where the height of the fins is fixed at 32 mm and has a thickness of 1.5 mm. The channel of the case of the solar panel acts as the airflow region is 50 mm height with a hydraulic diameter for the track is at 91.1504 mm. Two shapes were considered for the perforated and dimpled design; circular and triangle shapes. The cooling performance for these configurations is investigated based on simulated surface temperature reduction.

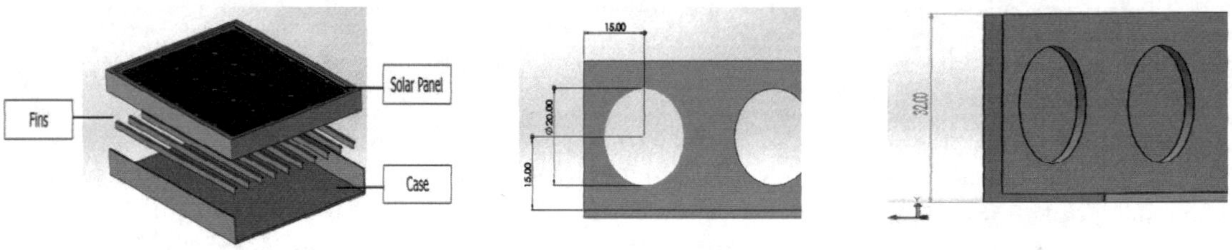

(a) (b) (c)

Figure 1. (a) Model of Solar PV with Fins (b) Perforated Circular Fins (c) Dimpled Circular Fins.

Table 1. Suitable Fin Dimension.

Parameter	Value
Thickness of Fin	1.5×10^{-3} m
Height of Fin	0.032 m
Spacing between Fins	0.018 m
Number of Fins	20/26 fins

2.2 CFD simulation

CFD modelling workflow can be classified by three distinct stages called pre-processing, processing, and post-processing. The computational domains were constructed during the pre-processing step, and the mesh was created for the solver. The processing stage included running all the predefined parameters and codes, including the characteristics and physical properties of the system, such as the properties of the fluid material, the physical properties of the solar panel and fin, and the boundary conditions. Finally, the post-processing phase presented the results graphically to enable visualisation and help understand and analyse the simulation results fully. Postprocessing data was then processed, analysed, and addressed.

The boundary conditions of the simulation were set as follows; a uniform velocity inlet was applied with a constant value of 4 m/s and 7 m/s at the inlet of cooling fins direction, and the pressure at the outlet was assigned to a continuous gauge pressure value of 101325 Pa. These speeds were selected based on available blowers used in validating the simulation result later. The external heat flux was applied to all the solar panel surfaces with a derived heat flux of 1175 W/m^2 at an ambient temperature of 35.1°C. Therefore, a parameter study was conducted according to the parameters shown in table 2 below. A Mesh independence study was performed to eliminate or reduce the number of cells and mesh size on the computational results. It is found that the solar panel temperature was stable, and there was no significant change in pressure distribution beyond the mesh size 8.0, with a total number of nodes was found at 4.1 million.

Table 2. Parameter Study Conditions.

Parameter	Forced Air Cooling		
Height of Fin (m)		0.032	
No. of fin		20	26
Perforation and Dimpled Density in a Fin		26	
Size of Perforation and Dimple in a Fin (m)		0.02	
The shape of Perforation in a Fin	None	Circular	Triangle
The shape of Dimple in a Fin	None	Circular	Triangle

2.3 Validation of CFD Study

Two methods were used to validate the CFD study; comparison of published simulation data [5,6] and experimental validation work. The experimental validation was varied in terms of maximum solar panel temperature and fin channel velocity outlet for better accuracy. The validation results for simulation and experimental works show excellent precision. The percentage difference in terms of the maximum temperature on the solar panel was 7.86% and 15.64%, respectively, while the percentage difference for the velocity outlet in experimental validation was 4.22%.

3. Results and discussion

The surface temperature of the solar PV at three main configurations; solid, perforated (Circular and Triangle) and Dimpled (Circular and Triangle), are as shown in figure 2 for air inlet velocity 4 m/s and figure 3 for air inlet velocity 7 m/s. A higher number of fins indicates better heat transfer and thus explains lower maximum and the average temperature in 26 fins configurations. The highest average temperature was simulated in a straight non-perforated/dimpled configuration at 51.79°C, a dramatic decrease in average temperature compared to a solar PV without a cooling fin. With the utilization of perforated/dimpled design, the average temperature shows signs of further temperature reduction with the lowest average temperature of 41.88°C on dimpled triangular configuration. Air velocity increment demonstrates better heat transfer and temperature reduction as the average temperature on all designs is slightly lower at higher velocity. The lowest average temperature simulated is 41.12°C on a perforated triangular configuration. It can be deduced based on figures 2 and 3; there is not much difference in temperature drop between circular and triangular shapes. This is because the simulated inlet speed is considerably high; the flow is in turbulent flow regardless of shapes. However, the perforated and dimpled configuration shows a slight difference in average temperature between 0.7% to 0.9%, with dimpled fins having the lower average temperature. The triangle shape also simulated the lowest maximum solar panel temperature compared to the circular shape. The triangle area destroys the thermal boundary layer area more frequently than the circular shape configurations [9, 7].

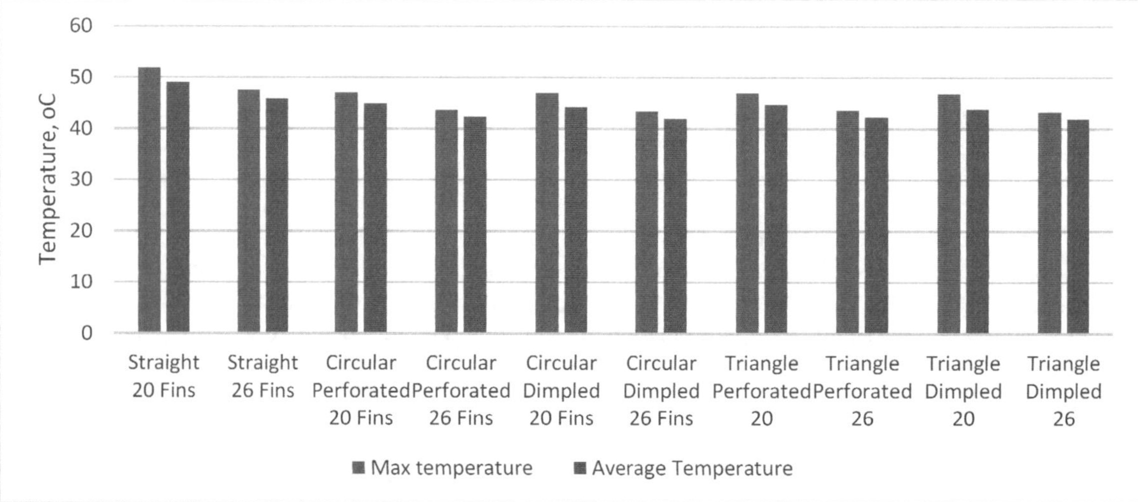

Figure 2. Simulated Solar PV Surface Temperature at 4m/s inlet velocity.

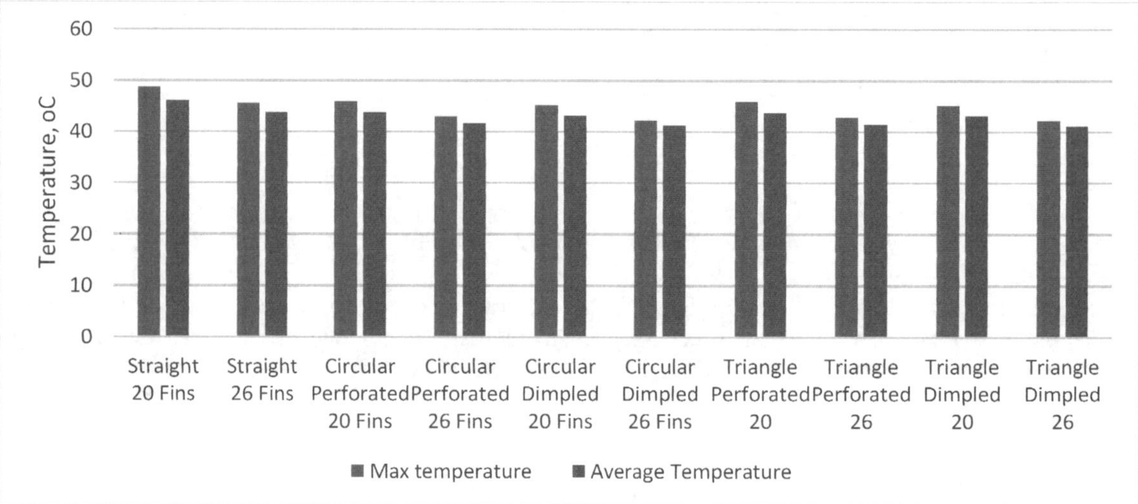

Figure 3. Simulated Solar PV Surface Temperature at 7 m/s inlet velocity.

The perforated and dimpled configuration also shows different behaviour of surface temperature, as shown in figure 4. Perforated fins allow airflow to go through the punched hole in the fins, promoting more turbulent flow. Thus, the cooling of the surface panel look asymmetric and might affect solar PV performance as the temperature gradient on the surface differs. On the other hand, dimpled fins keep airflow within its respected channel while the dimpled design causes the air to flow in turbulence flow.

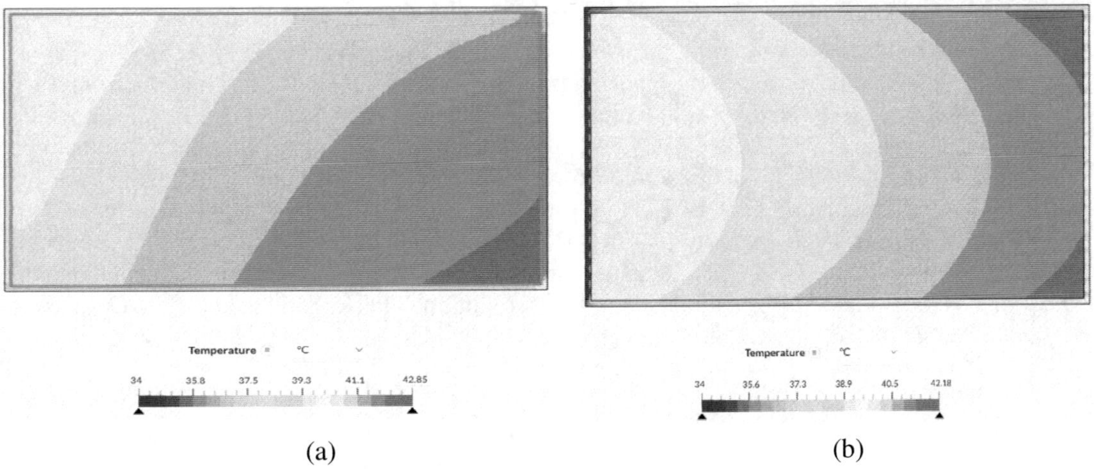

Figure 4. Panel Temperature Distribution with air inlet velocity at seven m/s on (a) Triangle Perforated Fin (b) Triangle Dimpled Fin.

4. Conclusion

In this paper, simulations were done using SimScale simulation software to study the effect of dimpled and perforated cooling fins on the surface temperature of solar PV. It is found that the use of perforated and dimpled fins manages further to reduce the average surface temperature of Solar PV. The lowest temperature simulated was at 41.12°C with a dimpled triangle shape configuration in the simulated study and condition. This study also observes no significant temperature difference between circular and triangular shape configuration, but; perforated and dimpled design does have temperature difference up to 9%. This might be attributed to the design of perforated and dimpled fins where the air is kept on its respective channels and having a more significant surface area for a better heat transfer rate. The number of fins and air inlet velocity also plays a positive role in reducing the surface temperature of the solar PV. Lastly, the design of the shapes can be further developed by applying other fin shape designs such as the trapezoidal and rectangular fin and operating at laminar inlet velocity. This is because these kinds of shape had a greater heat transfer rate than other shapes [10], and it is imperative to study the effect of these fin shapes on solar PV performances in terms of solar panel temperature, especially for passive cooling that haves typically a very low airspeed.

Acknowledgements

The authors wish to convey sincere gratitude to Universiti Malaysia Sabah (UMS) for supporting the current research work thru the FRGS research grant, project no. SDN0052 and GUG0331- 1/2019.

References

[1] Guramun S, Misaran M S, Ibrahim M K W and Rahman M M 2019 Trends of hybrid earth-air-pipe (EAP) photovoltaic cooling system for efficiency improvement: A review *J Mech. Eng. Res. Dev.* **42**(4) 191-5

[2] Haidar Z A, Orfi J and Kaneesamkandi Z 2018 Experimental investigation of evaporative cooling for enhancing photovoltaic panels efficiency *Results in Physics* **11** 690-7

[3] Saadi S, Benissaad S, Poncet S and Kabar Y 2018 Effective cooling of photovoltaic solar cells by inserting triangular ribs: a numerical study *Inter. J. Ener. and Environ. Eng.* **12**(7) 488-94

[4] Kim J and Nam Y 2019 Study on the cooling effect of attached fins on PV using CFD simulation *Energies* **12**(4) 758

[5] Arifin Z, Tjahjana D D D P, Hadi S, Rachmanto R A, Setyohandoko G and Sutanto B 2020 Numerical and experimental investigation of air cooling for photovoltaic panels using aluminum heat sinks *Int. J of Photoenergy*

[6] Popovici C G, Hudişteanu S V, Mateescu T D and Chereches N C 2016 Efficiency improvement of photovoltaic panels by using air cooled heat sinks *Energy Procedia* **85** 425–32
[7] Gupta S K and Thakur H 2015 Investigation to Compare Heat Augmentation from Plane, Dimpled, and Perforated Dimple Rectangular Fins Using Ansys *ELK Asia Pacific Journal*
[8] Raina G and Thakur N S 2019 Mathematical approach for optimizing heat sink for cooling of solar PV module *Int. J. Sci. Eng. Res* **7** 62-6
[9] Patil P V and Kumbhar D G 2014 CFD investigation on dimpled FINS with parameter variation for heat transfer augmentation *Int J Eng Res Technol (IJERT)* **3**(6) 662-6
[10] Shadlaghani A, Tavakoli M R, Farzaneh M and Salimpour M R 2016 Optimization of Triangular Fins With/Without Longitudinal Perforate for Thermal Performance Enhancement *J Mech. Sci. and Tech.* **30**(4) 1903-10

IOP Conf. Series: Materials Science and Engineering 1217 (2022) 012016 doi:10.1088/1757-899X/1217/1/012016

The effect of air velocity on the performance of the direct evaporative cooling system

Z Hassan, M S Misaran*, N J Siambun, M Adzrie

Faculty of Engineering, Universiti Malaysia Sabah (UMS), Jalan UMS, 88400 Kota Kinabalu, Sabah, Malaysia.

*Corresponding author: suffian@ums.edu.my

Abstract. This experimental study aimed to determine the effect of airflow velocity on the performance of a direct evaporative cooling system. Rectangular-shaped honeycomb cooling pads with a length of 34 cm, a width of 25 cm, and a thickness of 3.5 cm are used as cooling media. The main parameters of the study are low air velocity (2.3 ms^{-1}), medium (3.2 ms^{-1}), and high velocity (3.7 ms^{-1}). The data collected include dry bulb temperature, wet bulb temperature, output air temperature, input and output air velocity, input and output humidity, and solar radiation. These data are used to determine saturation efficiency, cooling capacity, temperature decreases, and feasibility index. The experimental results are presented in the form of tables and graphs and analysed based on existing theories. The results showed that the evaporative cooling system could produce output temperatures up to 27.5°C with input 31.4°C at low airspeed, 27.97°C with input 31.47°C at medium speed, and 27.70°C with input 31.30°C at high airspeed. It was concluded that a low airflow rate would add to the cooling efficiency, and the higher the airflow rate, the lower the cooling efficiency. The results showed that evaporative cooling is achievable with a feasibility index of $19.89 \leq F^* \leq 20.67$. The results also affirmed that cooling capability is higher where the feasibility indexes are comparatively low.

1. Introduction

The use of air conditioning nowadays is very advanced and used in various aspects of human life. Air conditioning for comfort is done by controlling temperature, humidity and cleaning the air to produce comfortable and clean air. With proper temperature, humidity, and even hygiene conditions, air conditioning can create a comfortable environment at low energy demand [1]. Energy is an essential factor in heating, ventilation, and air conditioning (HVAC) application as it is one of the main issues needed to be addressed sustainably.

Evaporative cooling technology is one of the air conditioners using the oldest principle, in which the air is cooled by the process of water evaporation [2]. Evaporative cooling can be called adiabatic saturation. The adiabatic Saturation device is a device that disperses air through a splash/spray of water [3]. There are three types of the evaporative cooling process is direct evaporative cooling, indirect evaporative cooling, and indirect-direct evaporative cooling [4]. The temperature, humidity, and cleanliness of the air provided, on the other hand, are more conducive to comfort.

Content from this work may be used under the terms of the Creative Commons Attribution 3.0 licence. Any further distribution of this work must maintain attribution to the author(s) and the title of the work, journal citation and DOI.
Published under licence by IOP Publishing Ltd

2. Direct evaporative cooling (DEC)

For the conditioning of room air, direct evaporative cooling is a standard process. The direct evaporative cooler (DEC) uses a fan to force warm air from outside through a porous pad onto which a continuous water spray is enabled to cool the air. When water evaporates from the moistened pad, the heat is absorbed by the water. Direct evaporative cooling has the disadvantage of cool only up to the inlet air's wet-bulb temperature. The sensible heat is converted to latent heat in this heat and mass exchange operation. As a result, direct evaporative cooling has two significant drawbacks, air retains moisture due to heat exchange with water, and temperature reduction is limited to the inlet air wet-bulb temperature.

Evaporation is the transformation of a liquid into water vapour when it is below its boiling point. At room temperature, water can evaporate by absorbing heat from the air. When the temperature of the environment or the water rises, the water evaporates more quickly. Evaporation takes place on the liquid's surface [5]. Numerous variables influence the evaporation mechanism, as shown in figure 1.

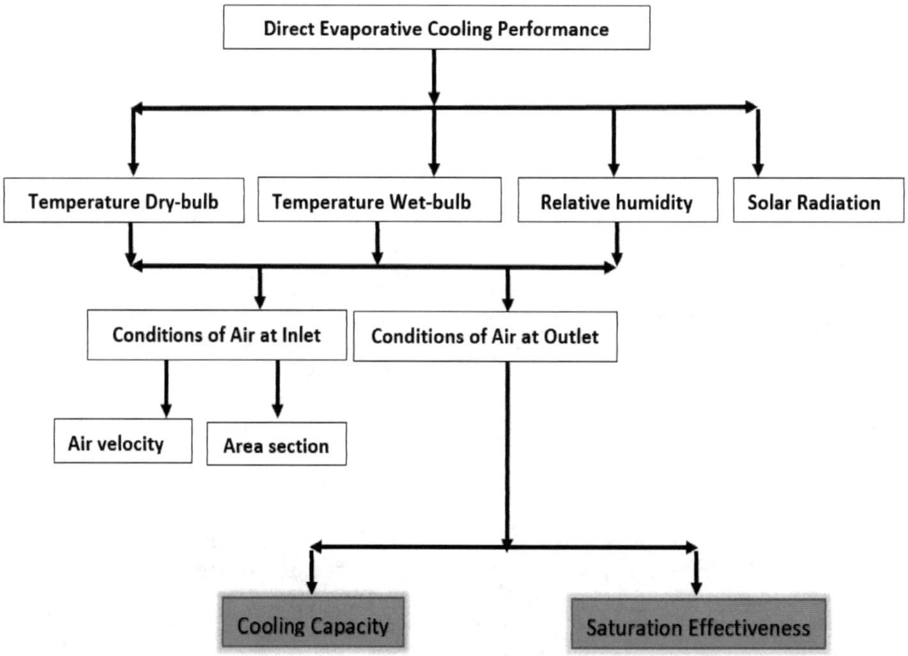

Figure 1. Main parameters of Direct Evaporative Cooling process.

The measurement of air velocity is just as critical as the other variables. The most important feature to consider when designing and operating these systems is the velocity of the air flowing through the pad (pad evaporative cooling systems). If this function is not selected correctly, the system's construction and maintenance costs will skyrocket. Evaporative cooling can increase air humidity, but it requires a continuous flow of outside air and the proper exhaust temperature [6]. Airflow not only has a cooling effect but can also cause discomfort in humans.

Air velocity refers to the rate at which air moves through space. The air is usually moved naturally, by convection or by fans. A device called an anemometer is used to measure air velocity. The anemometer is placed at the evaporative cooling unit's inlet to test the inlet air velocity. The mass flow rate is then calculated using the inlet air velocity, density, and the cross-section area of the inlet air duct [7]. There are various methods for raising air velocity, including increasing the fan's speed and size. X. She *et al.* [8] studied the airflow patterns of an evaporative cooling and dehumidification process for a hybrid refrigeration system. They discovered that they are useful for system design and optimization.

In an experiment set up following the principles of AMCA (Air Movement and Control Association) and ASHRAE, Koca *et al.* (1991) [9] evaluated the functionality of three separate cellulose pads

(American Society of Heating, Refrigerating and Air-Conditioning Engineers). The study's first pad had a thickness of 10 cm and a chamfer angle of 45 to 45 degrees, while the second pad had a thickness of 15 cm and a chamfer angle of 45 to 45 degrees, and the third pad had a thickness of 15 cm and a chamfer angle of 30 to 30 degrees. The study shows that at 1.5 to 2.5 ms^{-1} air velocity, the cooling efficiency of a pad with a 10 cm thickness and a 45°C chamfer angle yields 73% to 90%.
Cruz *et al*. (2006) [10] evaluated the efficiencies of three different pad materials at various temperatures and air velocities. The maximum cooling efficiency (80 % or more) was reached at 3.2 ms^{-1} air velocity at temperatures of 32 to 34°C in a study performed at four different temperature ranges with air velocity of 1.6, 3.2, 4.8, and 5.6 ms^{-1}, respectively.

Kocaturk and Yildiz (2006) [11] investigated some efficiency characteristics of a cellulose-based evaporative cooling pad at various air velocities. The air velocity through the cooling pad was set to 0.5, 1.0, and 2.0 ms^{-1}. The study reported that the cooling efficiency was 77% to 84 %, the temperature decreases between 6.7 and 5.6°C, and the amount of vaporized water was measured between 0.078 and 0.210 L min^{-1} m^{-2}.

Making an evaporative cooler that works in a hot, humid climate like Malaysia while maintaining human thermal comfort is a difficult task nowadays. Studies have shown that several factors are affecting the performance of evaporative coolers. One crucial factor would be air velocity variation to humidity. The most significant element that must be carefully determined throughout the design and operation of these devices is the velocity of the air moving through the pad. If this feature is not picked correctly, the system's construction and operation expenses will skyrocket. However, the remaining installation and operation costs have a negative impact on the firms' competitiveness.

Thus, this experimental study aimed to determine the effect of airflow velocity on the performance of a direct evaporative cooling system and help the people who design and operate these systems. The scope of study is to investigated the effect of air velocity in three air situations. In low air velocity (2.3 ms^{-1}), medium (3.2 ms^{-1}), and high velocity (3.7 ms^{-1}).

2.1 Methods and Experiment Setup

The study was carried out in Kolej Komuniti Kota Marudu, Sabah, at a location of 6.0367°N, 116.1186°E and the measurement was taken between 10:00 am and 01:00 pm in April 2021. The direct evaporative cooling (DEC) used for this study consisted of a 65-watt axial blower and a 10watt submersible water pump to circulate water in the system up to 1400 L/hour. The water is sprinkled on the upper side of the pad while air passes through the unit horizontally. A honeycomb cooling pad was used as cooling media with a length of 34 cm, a width of 25 cm and a thickness of 3.5 cm. Experiments were carried out to evaluate the performance of the direct evaporative cooling unit with different airflow rates.

Thermocouple's Type-K with an accuracy of ± 0.1°C was used to measure the air temperature. The relative humidity was measured using a humidity meter located at the inlet and outlet points of the evaporative cooling unit. At the same time, dry wet-bulb temperatures for ambient air were recorded using the sling psychrometer. An anemometer was installed at the inlet point of the evaporative cooling unit to measure the air velocity. The mass flow rate of air entering the evaporative cooling unit was determined by using the air velocity and the cross-section area of the inlet duct. A Solar Power Meter was used to record the solar radiation in the test location. To show that the weather condition is clear sky. The axial fan of the evaporative air cooler operates with three modes of speed: low, medium, and high. The water tank's water to the cooling pad is pumped by a submersible water pump located outside the water tank with a connecting pipe to the cooling tube. The schematic diagram and the experimental setup is shown in figure 2.

Figure 2. Diagram and photograph of a direct evaporative cooling and its pad.

Measurements were taken every 10 minutes for a total of 18 readings from all sensors throughout the 3-hour testing period. The air velocity was set using fan regulator at 2.3, 3.2, and 3.7 ms^{-1} and measurements were taken to compare the performance of DEC at different parameters. Measured inlet air velocities were used to calculate air mass flow rate in each test to achieve the same experimental conditions for results. Each reading was taken six times to ensure near-steady state conditions achieved and accurate readings. All the experiments were carried out in nearly identical conditions.

3. Performance evaluation

The evaporative cooling (EC) performances of the tested materials were evaluated based on the equations given below: The temperature drop of air (ΔT) across the EC pad is obtained with equation 1.

$$\Delta T = T_1 - T_2 \tag{1}$$

where T_1 for the inlet temperature and T_2 for outlet temperature.

The evaporative cooling saturation efficiency (ε) is calculated via equation 2 as described by ASHRAE (2001) [12]:

$$\varepsilon = (T_1 - T_2)/(T_1 - T_{wb}) \times 100 \tag{2}$$

where T_{wb} is evaporative inlet wet bulb temperature °C.

The mass flow rate of air across the EC pad should be measured to calculate the cooling capability. As shown in equation 1, it was determined by multiplying the measured air velocity at the outlet V by the average density of air, and the cross-section area (AC), of the material sample. The air mass flow rate is calculated using equation 3.

$$M_a = \rho VAC \tag{3}$$

where the sensible cooling capacity is calculated using the temperature drop and air mass flow rate as shown in equation 4.

$$Q_c = maC_p\,(T_1 - T_2) \tag{4}$$

Feasibility index (F*) is specified in equation 5

$$F^* = T_{wb} - (T_1 - T_{wb}) \tag{5}$$

where: Q_c is cooling capacity (kJ/h), ma for air flow rate (kg/s), C_p for specific heat of air (J/kgK).

A small F* number designates better feasibility for an evaporative cooling system. It denotes the ability of evaporative cooling to provide thermal comfort. The evaporative cooling effect increases as the temperature differential between the two rises. According to Watt, J. R 1963 [13], comfort cooling is when the F* is less than 10. When the F* is between 11 and 16, relative (lenitive) cooling is achieved, though F* > 16 is not recommended to use evaporative cooling systems [14]. The average daytime F* in Malaysia is 18, higher than the recommended margins.

4. Result and discussion

Experimental data were collected and analysed. The data presented in the form of temperature drop achieved, cooling efficiency, cooling capacity and feasibility index. As shown in table 1, the highest average temperature drop was measured at 2.3 ms^{-1}, at 4.21°C. While at a higher air velocity of 3.2 m/s and 3.7 m/s, the temperature decrease was almost identical to 3.5°C and 3.6°C, respectively. The highest temperature drop implies better average cooling efficiency at 76.66%, an approximate 10% better than the other velocity configuration. However, experimental data show higher air velocity institutes higher cooling capacity Q_c with the highest air velocity recorded 1.31 kW cooling capacity. This is expected as higher velocity means higher mass flow rates and more heat absorbed during the evaporation process. This result shows an exciting discovery where contact factors and air retention are essential in the cooling pad. A slower airflow through the cooling pad allows better surface contact on the pad surface and thus better heat transfer than higher velocity airflow. R.K. Kulkarni and S.P.S. Rajput (2010) [15] reported similar findings where the higher the air mass flow rate through the cooling surface, the less time the air stream is in contact with the wetted pad. As a result, less water is evaporated at the pad leads to lower relative humidity and temperature reduction in the space conditioned by the cooler. Similar studies show almost a similar trend, Timmos and Baughman (1984) [16], for example, calculated an average evaporative cooling saturation efficiency of 80% at 0.9 ms^{-1} air velocity in a poultry house with evaporative pad cooling in the South-east area of the United States. Other researchers achieve almost similar efficiency [17 – 20] at a range of 50% to 90%, which shows the achieved efficiency in the current DEC is acceptable. The average feasibility index was also analysed and showed impractical numbers, with the best Feasibility Index calculated at 19.89, higher than the preferred minimum index number, 18. One possible cause for this result is the high humidity recorded during the experiments; the relative humidity was relatively high at 70.8% - 72.5%, which is typical for a tropical climate.

Table 1. Performance of Direct Evaporative Cooling at a various air velocity.

Technical parameters	**Performance (Average)**		
	Low (2.3 m/s)	**Medium (3.2 m/s)**	**High (3.7 m/s)**
Dry-Bulb Inlet T_1 (°C)	31.72	31.47	31.30
Web-bulb Inlet T_{WB} (°C)	26.20	26.00	25.60
Dry-Bulb Outlet T_2 (°C)	27.5	27.97	27.70
Temperature Decreases (°C)	4.21	3.50	3.60
Cooling Efficiency (%)	76.66	64.24	63.82
Feasibility Index (F)	20.67	20.53	19.89
Cooling Capacity Q_C (KW)	0.95	1.30	1.31
Relative humidity RH_{in} (%)	70.8	72.32	72.50
Relative humidity RH_{out} (%)	73.05	74.98	75.85

The evaporative cooling saturation efficiency and capacity variations were also observed for an averaged single day, as shown in figures 3(a) and 3(b). The evaporative cooling saturation efficiency and capacity vary because of changes in relative humidity at the site of the experiment. Both data show the efficiency and cooling capacity were higher at the start of the experiments and increased until it reaches a maximum value before it gradually decreases towards the experiment's end, a left eschewed dome shape. This behaviour is co-related to the environment humidity of the experiment site, where the humidity is slightly lower at the beginning and becomes higher throughout the experiments.

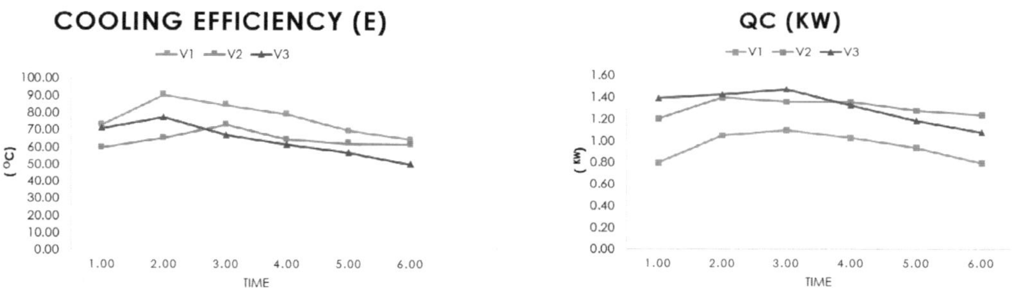

(a) (b)

Figure 3. Evaporative cooling saturation efficiency and Cooling Capacity Trends from 10 am to 1 pm.

5. Conclusion

In general, the evaporative cooling saturation efficiency of the direct evaporative cooler is influenced by air velocity, water flow rate, relative humidity, solar radiation, and the type of pad used. In this study, the air velocity variation was used to study the performance of the direct evaporative cooler and its feasibility in a Malaysian climate. It is found that air velocity does play an important role in increasing the cooling capacity of a direct evaporative system where the highest cooling capacity was measured at 1.31 kW at the highest air velocity. However, air velocity has an inverse effect on efficiency of the DEC where highest efficiency was measured at lowest velocity configuration at 76.66%. Slow air velocity promotes better retention time of air on the cooling pad surface and allows more evaporation process took place and reduces temperature. In the future, more study needs to be done to design better cooling pad that allows better efficiency at higher velocity.

Acknowledgements

The authors wish to convey sincere gratitude to Universiti Malaysia Sabah (UMS) for supporting the current research work thru UMS research grant, project no. GA19100.

References

[1] Misaran M S, Sing C F and Adzrie M 2019 Study of wyee-tee duck design at various protrusion and guide vane location using CFD *CFD Letters* **11**(11) 39-47

[2] Jain J K and Hindoliya D A 2011 Experimental performance of new evaporative cooling pad materials *Sustainable Cities and Society* **1**(4) 252-6

[3] Stoecker W F and Jones J W 1982 *Refrigeration and Air Conditioning* New York: The McGaw-Hill

[4] Wang S K and Wang S K 2000 *Handbook of air conditioning and refrigeration* New York: McGraw-Hill

[5] Liberty J T, Ugwuishiwu B O, Pukuma S A and Odo C E 2013 Principles and application of evaporative cooling systems for fruits and vegetables preservation. *Int. J. Current Eng. and Tech.* **3**(3) 1000-6

[6] Bhatia A Principles of Evaporative Cooling System [Online], [Accessed 22.02. 2014]. Available from World Wide Web: www. pdhonline. org/courses/m231/m231content. pdf. 2012.

[7] Mohammada A T, Mata S B, Sulaimana M Y, Sopiana K and Al-abidia A A 2013 Experimental Performance of a Direct Evaporative Cooler Operating in Kuala Lumpur *Int. J. Thermal & Env. Eng.* **6**(1) 15-20

[8] She X, Yin Y, Dong Y and Zhang X 2016 Investigation on air flow patterns of evaporative cooling and dehumidification process for a hybrid refrigeration system *App. Thermal Eng.***95** 79-94
[9] Koca R W, Hughes W C and Christianson L L 1991 Evaporative cooling pads: test procedure and evaluation *App.Eng. in Agriculture* **7**(4) 485-90
[10] Cruz V F, Perissinotto M, Lucas E M and De Moura D J 2006 Efficiency evaluation of pad cooling systems design with different pad materials. *In Proc. XVI. CIGR World Congress* 403-404
[11] Kocaturk U and Yildiz Y 2006 Some Performance Characteristics of Evaporative Cooling Pad At Three Air Velocities in Cukurova Region *J. Agric. Fac.* **21**(4) 101-8
[12] Rawangkul R, Khedari J, Hirunlabh J and Zeghmati B 2008 Performance analysis of a new sustainable evaporative cooling pad made from coconut coir *Int. J. Sust. Eng,* **1**(2) 117-31
[13] Watt J 2012 *Evaporative air conditioning handbook*. Springer Science & Business Media
[14] Saw I J and Fong O R 2019 Effects and Usefulness of Evaporative Cooling in Malaysia *Malaysia Green Building Council*
[15] Kulkarni R K and Rajput S P Comparative performance analysis of evaporative cooling pads of alternative configurations and materials 2013 *Int. J. Adv. Eng. & Tech.* **6**(4) 1524.
[16] Timmons M B, Baughman G R 1984 A plenum concept applied to evaporative pad cooling for broiler housing *Trans. of the ASAE* **27**(6) 1877-81
[17] Timmons M B, Baughman G R, Parkhurst C R 1981 Use of evaporative cooling to reduce poultry heat stress *Microfiche collection*
[18] Ganie A A, Kaur D, Nagra S S, Chopra S, Nayyar S 2018 Effectiveness of Various Cooling Systems during High Environmental Temperature on Production and Welfare Indices of Laying Pullets in Cage System of Rearing *J. Animal Research* **8**(5) 813-21
[19] Dağtekin M E, Karaca C, Yildiz Y I 2009 Long axis heat distribution in a tunnel-ventilated broiler house equipped with an evaporative pad cooling system *Animal Production Science* **49**(12) 1125-31
[20] Kulkarni R K, Rajput S P 2010 Theoretical performance analysis of jute fiber rope bank as media in evaporative coolers *Indian J. Sci. and Tech.* **3**(10) 1075-80

IOP Conf. Series: Materials Science and Engineering 1217 (2022) 012017 doi:10.1088/1757-899X/1217/1/012017

In situ poly(melamine-formaldehyde) microencapsulation of diglycidyl ether of bisphenol-A epoxy and pentaerythritol tetrakis(3-mercaptopropionate) for self-healing applications

H Ghazali[1,2]*, K Ghazali[3], and R Yusoff[3]

[1] Centre for Advanced Composite Materials (CACM), Universiti Teknologi Malaysia, 81310 Johor Bahru, Johor, Malaysia
[2] School of Mechanical Engineering, Universiti Teknologi Malaysia, 81310 Johor Bahru, Johor, Malaysia
[3] Faculty of Science and Natural Resources, Universiti Malaysia Sabah, 88400 Kota Kinabalu, Sabah, Malaysia.

*Corresponding author: habibahghazali@utm.my

Abstract. In this study, microcapsules as potential candidates for self-healing agents were prepared by *in situ* polymerisation, taking place in oil-in-water emulsion. Poly(melamine-formaldehyde) is employed as shell material and diglycidyl ether of bisphenol A as polymerisable core materials and Pentaerythritol Tetrakis(3-Mercaptopropionate) as its hardener. The geometry, shell features, size distributions, core content, and the reactivity of the microcapsules were studied by scanning electron microscopy (SEM), optical microscopy (OM), Soxhlet extraction method and differential scanning calorimetry (DSC). Microcapsules with different sizes and distributions were obtained by adjusting the stirring speeds during the preparation stage. From the results, it was established that the spherical microcapsules fabricated using this technique, resulted in satisfactory size and shell structure with shell thickness of less than 2 μm. The microcapsules possess high core content at about 90 wt.% for each size range of microcapsules and it was also observed that the viscosity of the core content decreased at an elevated temperature. The results obtained in this work indicate that these microcapsules possess the characteristics that can be potentially used for self-healing applications.

1. Introduction

Microcapsules are near-spherical particles consist of encapsulating brittle material that protects the core material from the exterior which only to be released at specific conditions when necessary [1, 2]. The shell of a microcapsule is expected to release its content when it is broken by pressure, melting by heating, or dissolved in a solvent. In capsule-based self-healing material approach, the microcapsule is designed to store the healing agent that releases its content when triggered by damages such as microcracks that cracks of open its shell [3, 4].

The capsule-based self-healing approach was initially introduced by White *et al.* [3] and has since evolved. In their study, they have introduced the capsule-catalyst healing system as schematically

Content from this work may be used under the terms of the Creative Commons Attribution 3.0 licence. Any further distribution of this work must maintain attribution to the author(s) and the title of the work, journal citation and DOI.
Published under licence by IOP Publishing Ltd

illustrates in figure 1(a). This system involves the curing reaction between encapsulated dicyclopentadiene (DCPD) monomer with the Grubbs' catalyst that were pre-dispersed inside the host polymeric material. Another model of capsule-catalyst system was presented by Kamphaus *et al.* [5]; which demonstrated the curing reaction of encapsulated DCPD with pre-dispersed tungsten (VI) chloride (WCl_6) catalyst.

Another type of capsule-based self-healing approach involves the microencapsulation of both polymerisable monomer and its liquid hardener inside brittle protective shell as illustrates in figure 1(b). When damage such as microcrack happen to the host material, as the crack propagates it will rupture both capsules to release their contents. The polymerisable monomer will start the cure reaction once it is in contact with the curing agent and hold the crack surface together. Yuan *et al.* [6], have investigated a dual-capsule system containing diglycidyl tetrahydro-o-phthalate (DTHP) epoxy resin and its curing agent, where they have reported 100 % recovery of the host polymeric material original properties after healing. Ye *et al.* [7] presented an ultra-fast dual-capsule healing chemistry consisting encapsulated epoxy and pentafluoride–ethanol complex (SbF_5) in their study. They have reported 60% repair of impact damage within a few seconds with this healing chemistry.

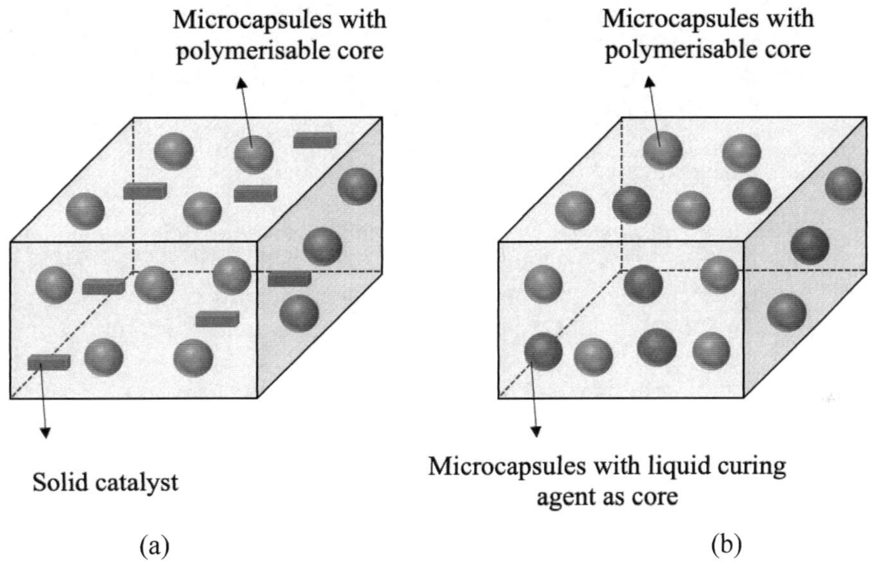

Figure 1. Microcapsules based self-healing approaches: (left) capsule-catalyst (middle) dual-capsule (right) single-capsule.

To be potentially employed in self-healing applications, there are several requirements that must be met by these microcapsules. Firstly, the core materials must possess low viscosity and rather high core weight percentage so that they can easily flow to the crack surfaces and enough to fill up the void [8, 9]. Apart from that, the core content of microcapsules should have high stability and remain reactive for a long time [6]. In the case of an elevated temperature applications the microcapsules must be able to tolerate high temperature, therefore, the core content with high thermal stability is preferred [6, 10]. And finally, the shell of the microcapsules must be brittle enough to rupture when required in order to release the healing agent [11, 12].

In the previously reported microencapsulated self-healing systems, epoxy resins such as DTHP [6] and a mixture of diglycidyl ether of bisphenol A (DGEBA) and diglycidyl ether of resorcin (DGRE) [13] have been used as polymerisable monomer cores due to their low viscosity. However, these epoxy resins possess lower thermal stability in comparison to the unmodified DGEBA which limits their potential to be used as self-healing agent in an elevated temperature application. The aim of the current study is to synthesis the microcapsules containing the unmodified DGEBA and its hardener through *in situ* polymerisation and characterise their properties. In this study the unmodified DGEBA epoxy and Pentaerythritol tetrakis(3-mercaptopropionate) (PETMP) as its curing agent were encapsulated by an *in situ* polymerisation technique with poly(melamine-formaldehyde) (PMF) as a shell material. The initiation of healing agent polymerisation is achieved by the interaction of epoxy with its hardener on the crack plane. Viscosity of the core materials, the core content weight percentage, the reactivity of the contents, the morphology and the microcapsules size distribution were studied and discussed in detail.

2. Experimental

2.1. Materials
A diglycidyl-ether of bisphenol-A (DGEBA) epoxy resin – Araldite-F supplied by Ciba-Geigy, Australia, was employed as the polymerisable core of this self-healing system. This epoxy resin has an epoxy content of 5.20 – 5.35 equiv/kg. The Pentaerythritol tetrakis(3-mercaptopropionate) (PETMP), was used as a hardener while the 2,4,6-Tris (dimethylaminomethyl)phenol (DMP-30) was used as catalyst. Melamine with purity of 99% and formaldehyde (36.5 – 38 wt.% in H_2O) were used as shell-forming monomers. Triethanolamine and citric acid were used to control the pH value of the emulsified solution. 1-octanol was used to eliminate the bubble of the emulsified solution containing healing agents. For the preparation of emulsifier solution, styrene, maleic anhydride, benzoyl peroxide (BPO), toluene, sodium hydroxide (NaOH) were used. All substances other than the DGEBA were purchased from Sigma-Aldrich and used without further purification.

2.2. Preparation of the microcapsules

2.2.1. Preparation emulsifier solution. Emulsifier used in this study is the poly(styrene–maleic sodium) (PSMS*)* solution and the first stage in preparing this emulsifier is to prepare the styrene-maleic anhydride (SMA) copolymer. Firstly, a mixture of 60 g maleic anhydride, 1.2 g benzoyl peroxide and 24 ml styrene dissolved in 500 ml toluene were prepared in a three neck round bottom flask (1000 mL) equipped with reflux condenser and a nitrogen gas inlet system. The mixture was stirred at 75°C at 100 rpm inside a silicone oil bath until its cloud point. After reaching its cloud point, a reagent consist of toluene (260 ml) and styrene (46.4 ml) was added slowly (i.e.: drop-wise) into the mixture with the aid of dropping funnel. The temperature was then increased to 80°C and maintained at this temperature for 1 hour before cooled to room temperature.

Using filtration process, the SMA copolymer was recovered from the mixture reaction as white fine particles. The particles were washed repeatedly with toluene and oven dried at 60°C for 24 hours to a constant weight. The 10 wt. % hydrolysed emulsifier solution was then prepared by dissolving SMA copolymer particles in sodium hydroxide solution (10% NaOH in H_2O) and stirred at 80°C for 10 hours (until solution become clear). It must be cooled down to room temperature before subsequent usage.

2.2.2. Preparation of melamine-formaldehyde prepolymer. The prepolymer of melamine (22.4 g) and formaldehyde (48.4 g) was synthesised at 70°C for 13 minutes. The mixture was stirred continuously with magnetic agitator and 2 drops of triethanolamine were added to the solution to adjust the pH value to about 9-10. The solution was then cooled down to 50°C. The prepolymer was used directly in the following synthesis.

2.2.3. Preparation of DGEBA-loaded microcapsules. The microencapsulation of the DGEBA begins by adding 320 grams of water into 80 grams 10 wt. % of PSMS emulsifier solution in the three necks round flask. The mixture was stirred for about 2 minutes and then two drops of 1-octanol were added to eliminate surface bubbles of the mixture. The pH value of the mixture was controlled by adding 3 wt. % citric acid into the mixture. Once the pH was in the appropriate range (pH 4.2), DGEBA epoxy (100 g) was added into the mixture. Polytetrafluoroethylene (PTFE) stirring blades attached to an overhead stirrer were lowered to approximately halfway from the surface of the solution. The solution then stirred

IOP Conf. Series: Materials Science and Engineering 1217 (2022) 012017 doi:10.1088/1757-899X/1217/1/012017

at 60°C for 2 hours with different level of stirring speeds (450 rpm and 1200 rpm) to study the effect of the agitation rate. Stirring was done without any excessive vibration.

Subsequently, the prepolymer of melamine-formaldehyde solution was added into the DGEBA epoxy emulsion with continuous agitation for 30 minutes before adding 35 ml of 3 wt.% citric acid to adjust and maintain the pH value at about 3. The speed of agitation was reduced to 350 rpm (for synthesis with lower stirring speed) and 1100 rpm (for synthesis with higher stirring speed) with the addition of citric acid for about 2 hours. Next, the temperature of the system was increased to 70°C and the speed was again reduced to 260 rpm with continuous agitation for 2 hours.

Eventually, the reaction mixture was cooled down to room temperature and the resultant slurry of microcapsules was sieved using mesh sieve with different aperture opening with running water. Microcapsules of the desired size were collected from appropriate sieve. The clumps of microcapsules were dried at room temperature for 24 hours and then placed inside the drying cabinet for another 24 hours. Figure 2(a) shows the experimental setup used in the preparation of the microcapsules.

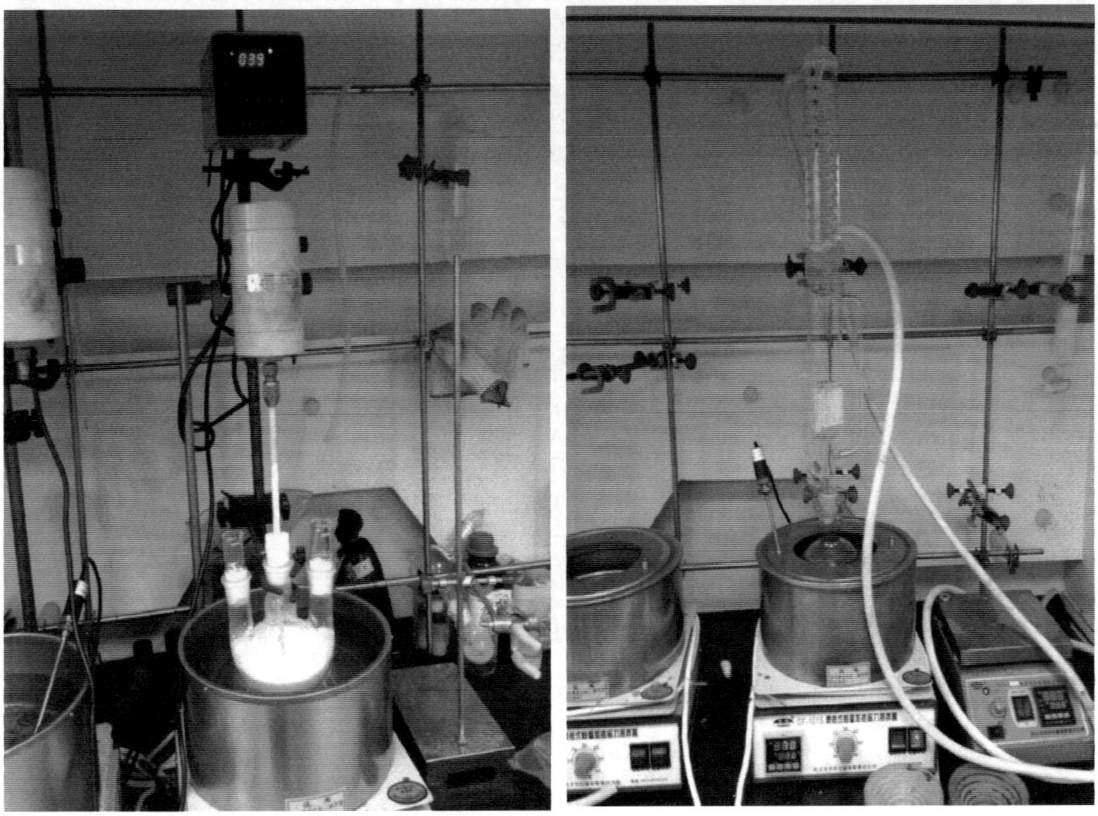

Figure 2. (a) Experimental setup equipped with an overhead stirrer and PTFE blade used in the preparation of the microcapsules (b) Soxhlet extractor used to determine the core content.

2.2.4. Preparation of PETMP-loaded microcapsules. Microcapsules containing PETMP were prepared in two steps. Firstly, PETMP was encapsulated by adapting a similar process to make DGEBA epoxy-loaded microcapsules. Secondly by adapting procedure established by Yuan *et al.* [14], the resulting microcapsules were then dispersed into the catalyst solution, DMP-30, at 40°C for 48 hours and then filtrated and washed with ethyl ether. The microcapsules were then left to dry at room temperature for a few minutes in a covered container to avoid evaporation of DMP-30 and immediately used in the fabrication of self-healing material afterward.

2.3 Microcapsules characterisations

Microcapsules size distribution was calculated using image analysis on microscopic images obtained by optical microscope (Leica, DMRXE). The samples of microcapsules were taken from 5 different batches for both lower and higher stirring speeds. From here onwards, microcapsules obtained from the lower stirring speed is called Type-1 and those obtained from higher stirring speed is called Type-2. The geometry, surface features and shell thickness of the microcapsules were observed using Scanning Electron Microscopy (SEM Hitachi S4600) and (SEM Zeiss EVO/Qemscan). Specifically for the observation of the shell thickness, prior to SEM observation, microcapsules were dispersed on double-sided carbon tape and crushed with a razor blade to release the core content. The capsules were then dried in a vacuum chamber for 16 hours. This is done to protect the SEM's vacuum system from contamination. After that, the samples were coated with gold using sputter coater to improve the imaging of the samples.

The Soxhlet extraction method was employed to determine the core content of the microcapsules [6, 15]. Soxhlet extractor (figure 2(b)) was used with acetone as an extraction solvent. A small number of microcapsules was thoroughly ground to ensure all the capsules were broken. The microcapsule samples were then wrapped using the Whatman™ qualitative filter paper (Grade 3, 6 μm, medium flow) and weighed. This is the total weight of the core content, W_c, and the shell, W_s, of the microcapsules, W_M. The microcapsule samples were extracted by Soxhlet apparatus for 240 hours to remove the core content. After extraction, the weight of the sample was measured once again. The final weight is the weight of the shell, W_s. The extraction was done 3 times for both DGEBA-loaded and PETMP-loaded microcapsules synthesised with different stirring speeds to get the average value.

The viscosity of both core ingredients, DGEBA epoxy and the PETMP used in this study were measured using a parallel-plate type rheometer (Anton Paar - MCR 300 rheometer) at two different temperatures, 25°C and 70°C.

Differential Scanning Calorimetry was used to investigate the reactivity of the DGEBA and the PETMP. Two samples were prepared; 1) unencapsulated epoxy resin mixed its unencapsulated hardener and 2) crushed microencapsulated epoxy resin mixed with its crushed microencapsulated hardener. Calorimetric analysis was carried out by a DSC, TA Instruments (2920 Modulated DSC). The analysis was performed by scanning the temperature from 30°C to 200°C at a scan rate of 10°C/min. The mass of the sample was about 10 mg, in an aluminium pan with a diameter of 5 mm.

3. Results and discussion

3.1. Effect of stirring speed

Figure 3 shows an example of the microscopic image used in the determination of the size distribution of the microcapsules. In this study, it was found that the microcapsules with two different size ranges can be obtained by controlling the stirring speed, similar to previously reported studies [9, 16]. At a higher stirring speed (1200 rpm), microcapsules with a smaller size range can be obtained, while microcapsules with a larger size range can be obtained by lowering the stirring speed (450 rpm). The higher agitation speed shows the breaking down of the DGEBA epoxy into smaller droplets leading to smaller size microcapsules. The average larger size (Type-1) of the DGEBA-loaded microcapsules is 110.6 μm and for PETMP-loaded microcapsules is 136.4 μm. For smaller sized capsules (Type-2), the average size was about 65.2 μm for DGEBA-loaded microcapsules and 66.1 μm for PETMP-loaded microcapsules. Figure 4 shows the size distribution of microcapsules synthesised with two different stirring speeds in this study. As can be seen in this figure, not only the lower stirring speed yields a larger average size microcapsules it also produced a wider particle size distribution in comparison to the microcapsules size distribution with the higher stirring speed.

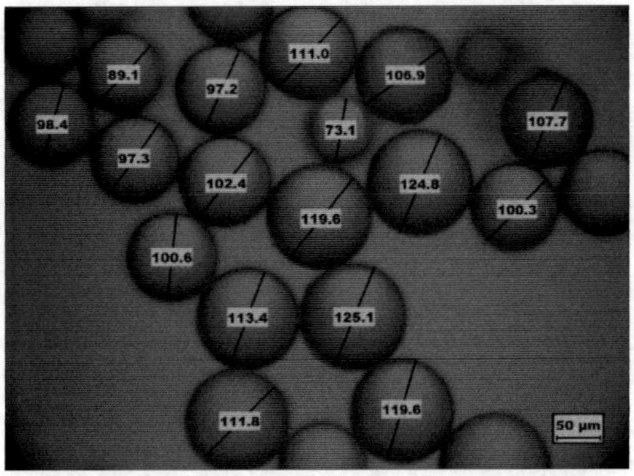

Figure 3. Example of microscopic image used in the determination of capsules size distribution with optical microscopy image.

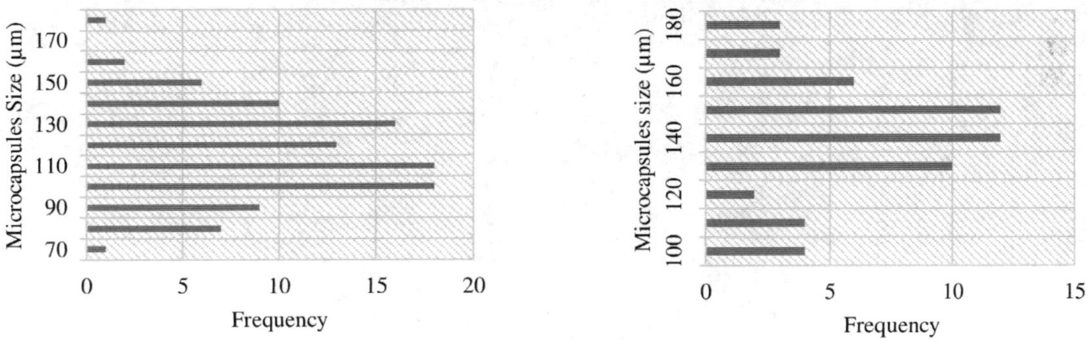

(a)

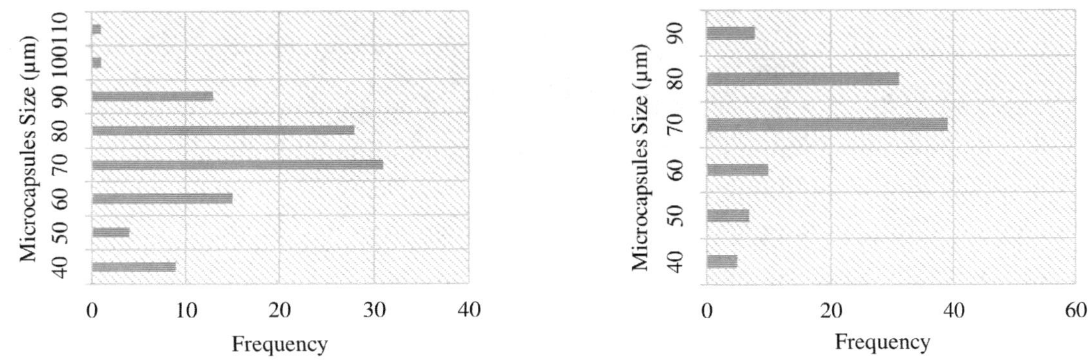

(b)

Figure 4. Size distribution of microcapsule (a) Type-1 microcapsules (left) DGEBA epoxy-loaded microcapsules (right) PETMP-loaded microcapsules (b) Type-2 microcapsules (left) DGEBA epoxy-loaded microcapsules (right) PETMP-loaded microcapsules.

3.2. Microcapsule's geometry and shell feature

Figure 5(a) and 5(b) shows the geometry and feature of the microcapsules obtained by scanning electron microscope (SEM) for both DGEBA-loaded and the PETMP-loaded microcapsules respectively. As observed from these figures, all microcapsules have similar spherical features. Both DGEBA-loaded and PETMP-loaded microcapsules appear to have nonporous surfaces resulting from the deposition of the low-molecular-weight PMF prepolymer at the epoxy droplet-water interface [17]. The nonporous surface will provide the microcapsules a long-shelf life. The shell surface of the DGEBA-loaded microcapsules appears very smooth while the shell of PETMP-loaded microcapsule shows the appearance of powder-like nano PMF debris, which is the shell material itself. This condition was reported to be helpful as it will improve the locking between the microcapsules with the host material and subsequently improve the healing efficiency [18]. It was observed that the shell thickness of the DGEBA epoxy-loaded microcapsules was 1.48 (±0.20) µm, and for PETMP-loaded microcapsules the shell thickness was 1.53 (±0.24) µm as shown in figure 5(c). These thicknesses are within the range of self-healing microcapsules synthesised in the earlier studies [19-21], therefore concluded to be potentially viable for self-healing applications.

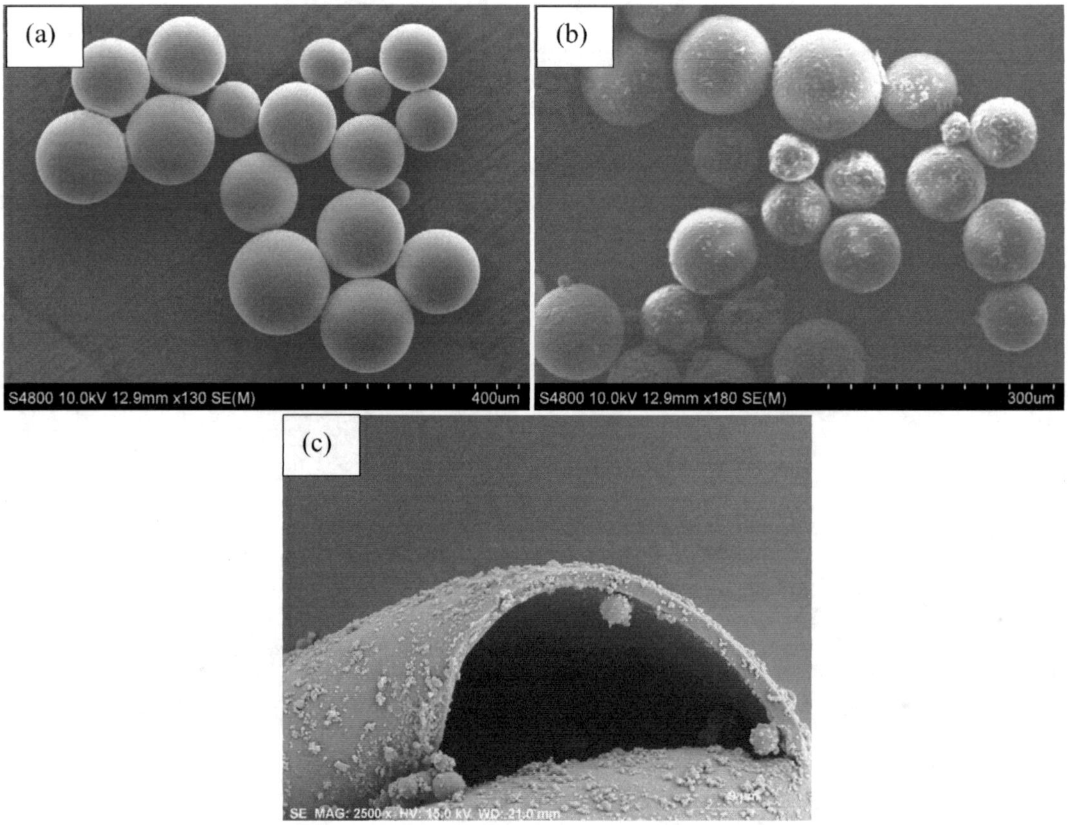

Figure 5. (a) DGEBA-loaded microcapsules, (b) PETMP-loaded (c) microcapsules shell thickness.

3.3. Microcapsules core content and viscosity

The data for the core content of the microcapsules, is presented in table 1. The estimation of the core content percentage is the calculated using equation (1):

$$W_c\ (\%)=\left(1-\frac{W_S}{W_M}\right)\times 100 \quad (1)$$

where W_c is the core content, W_s is the weight of the shell, and W_M is the total weight of the microcapsules.

The extremely high content of core inside both types of microcapsules ensures that enough healing agent will be able to be released and fill in gap and caused by the microcrack.

Table 1. Core content percentage of both Type-1 and Type-2 microcapsules.

Microcapsules	Core Content (%)
DGEBA Type-1	96.70
PETMP Type-1	93.10 (0.6 wt.% DMP-30)
DGEBA Type-2	90.29
PETMP Type-2	89.63 (0.6 wt.% DMP 30)

Density of DGEBA, PETMP and PMF are 1.24 g/cm³, 1.28 g/cm³ and, 1.10 g/cm³ respectively

Figure 6 shows that the viscosity of DGEBA epoxy and PETMP decreases with increasing temperature. At room temperature (25°C) the viscosity of DGEBA epoxy is at about 10 Pa.s, and the viscosity drop to 0.1 Pa.s at 70°C. This is comparable to the viscosity of DTHP discuss in [6]. As for the PETMP the viscosity does not show huge different, where at both temperatures the viscosity of PETMP remains below 1 Pa.s. This indicates that better flow ability can be achieved when these microcapsules used as self-healant at an elevated temperature.

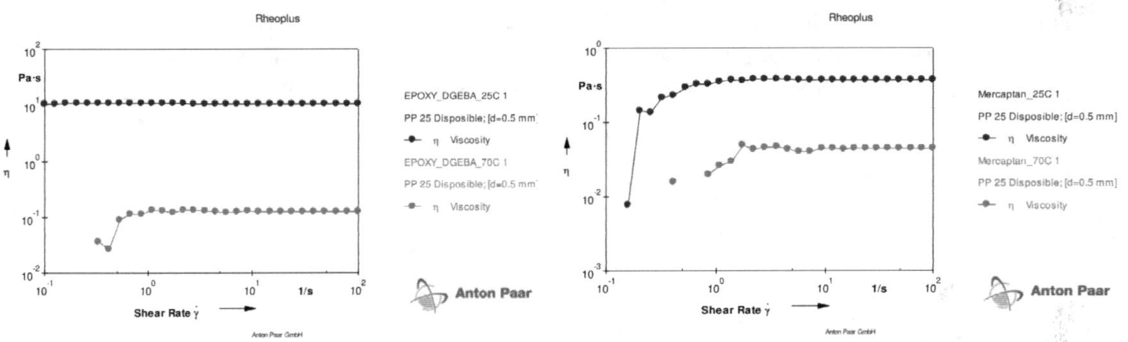

Figure 6. Shear-rate-dependent viscosity curves the microcapsules core content (left) DGEBA (right) PETMP. Blue curves show viscosity at room temperature (25°) whole the red curves show viscosity at and elevated temperature (70°).

3.4. Reactivity of microcapsules

The reactivity of the dual-capsule self-healing agent was analysed with differential scanning calorimetry (DSC). In this study, 1 to 1 stoichiometry between DGEBA and PETMP was decided. At 1:1 DGEBA/PETMP weight ratio, there is an excess of DGEBA compared to PETMP groups. However, with DMP-30 as catalyst, this will cause the DGEBA to self-cure. Evidently, variation greatly from 1:1 ratio will cause problems of incomplete cure [22].

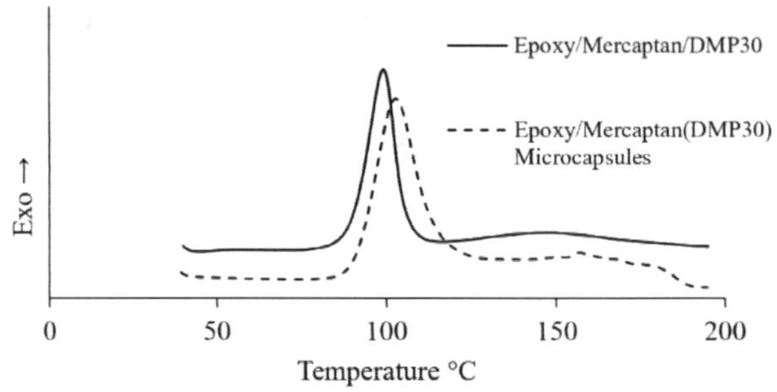

Figure 7. DSC analysis showing reactivity of DGEBA and PETMP as self-healing agent (solid line – mixture of unencapsulated DGEBA resin/PETMP/DMP-30, dashed line – mixture of ground DGEBA/PETMP microcapsules).

At first, a mixture of unencapsulated healant (DGEBA epoxy resin, PETMP, DMP-30 with weight ratio of 100:100:2) was heated from 30°C to 200°C at a rate of 10°C/min in N_2 environment. An obvious exothermic reaction peak can be observed (figure 7 – solid line) due to the cure reaction of the mixture. The exothermic peak of this curing is located at approximately 98°C, signifying that the fastest cure for this mixture will happen at this temperature. The onset of the cure peak is about 70°C. Second analysis was done on grounded DGEBA- and PETMP-loaded microcapsules (readily infused with DMP-30) with a weight ratio of 100:100. The exothermic peak shifted a little (figure 7 – dashed line), as there is no way to ensure 100% of the ground microcapsules were broken; therefore, the stoichiometric ratio of DGEBA and PETMP in the mixture might not be the same as the unencapsulated healing agents.

4. Conclusion

In this work, microencapsulation of DGEBA epoxy and its hardener PETMP was done by *in situ* polymerisation of poly(melamine formaldehyde) in an oil-in-water emulsion technique. Microcapsules with two different size ranges were prepared by controlling the agitation rate (ranging from 460 – 1200 rpm). All microcapsules featured a perfect spherical shape, and shell morphology of DGEBA epoxy-loaded microcapsules shows a smooth surface, whereas for PETMP-loaded microcapsules, nano-size PMF debris can be observed. The shell thickness of DGEBA DGEBA-loaded microcapsules is 1.48 (±0.20) μm and 1.53 (±0.24) μm for PETMP-loaded microcapsules. The core content for all microcapsules fabricated was approximately ≈ 90 wt. % and the viscosity of the core content can be seen decreased as the temperature increase. The microcapsules synthesised in this work can be concluded to be potentially used as self-healing agents for polymer composites material used under elevated temperature service condition.

Acknowledgement

The author(s) disclosed receipt of the following financial support for the research, authorship, and/or publication of this article: This work was supported by Universiti Teknologi Malaysia, UTMER grant, cost center no. Q.J130000.3851.19J39.

References

[1] Ma G and Su Z-G 2013 *Microspheres and microcapsules in biotechnology: design, preparation and applications* (Singapore: CRC Press)
[2] Lang S and Zhou Q 2017 Synthesis and characterization of poly(urea-formaldehyde) microcapsules containing linseed oil for self-healing coating development *Progress in organic coatings* **105** 99-110
[3] White S R, Sottos N R, Geubelle P H, Moore J S, Kessler M R, Sriram S R, Brown E N, and Viswanathan S 2001 Autonomic healing of polymer composites *Nature* **409**(6822) 794-7

[4] Gao C, Ruan H, Yang C and Wang F 2021 Investigation on microcapsule self-healing mechanism of polymer matrix composites based on numerical simulation *Polymer composites* **42**(7) 3619-31

[5] Kamphaus J M, Rule J D, Moore J S, Sottos N R and White S R 2008 A new self-healing epoxy with tungsten (VI) chloride catalyst *J The Royal Soc. Interface* **5(18)** 95-103

[6] Yuan Y C, Rong M Z, Zhang M Q, Chen J, Yang G C and Li XM 2008 Self-Healing Polymeric Materials Using Epoxy/Mercaptan as the Healant *Macromolecules* **41**(14) 5197-202

[7] Ye X J, Zhang J-L, Zhu Y, Rong M Z, Zhang M Q, Song Y X and Zhang H X 2014 Ultrafast Self-Healing of Polymer toward Strength Restoration *ACS App. Mat. & Interfaces* **6**(5) 3661-70

[8] Brown E, Sottos N R and White S R 2002 Fracture Testing of a Self-Healing Polymer Composite *Experimental Mechanics* **42**(4) 372-9

[9] Zhang M Q and Rong M Z 2011 *Self-healing polymers and polymer composites* (New Jersey: John Wiley & Sons)

[10] Yuan L, Liang G, Xie J, Li L and Guo J 2006 Preparation and characterization of poly (urea-formaldehyde) microcapsules filled with epoxy resins *Polymer* **47**(15) 5338-49

[11] Liu X, Sheng X, Lee J K and Kessler M R 2009 Synthesis and Characterization of Melamine-Urea-Formaldehyde Microcapsules Containing ENB-Based Self-Healing Agents *Macromolecular Mat. and Eng.* **294**(6-7) 389-95

[12] Murphy E B and Wudl F 2010 The world of smart healable materials *Progress in Polymer Sci.* **35**(1) 223-51

[13] Yuan Y C, Ye Y, Rong M Z, Chen H, Wu J, Zhang M Q, Qin S X, and Yang G C 2010 Self-healing of low-velocity impact damage in glass fabric/epoxy composites using an epoxy–mercaptan healing agent *Smart Materials and Structures* **20**(1) 015024

[14] Yuan Y C, Rong M Z, Zhang M Q and Yang G C 2009 Study of factors related to performance improvement of self-healing epoxy based on dual encapsulated healant *Polymer* **50**(24) 5771-81

[15] Yuan Y C, Rong M Z and Zhang M Q 2008 Preparation and characterization of microencapsulated polythiol *Polymer* 49(10) 2531-41

[16] Qiao L, Xue Y and Zhang Q 2018 Synthesis and characterization of phenol–formaldehyde microcapsules for self-healing coatings *J Mat. Sci.* **53**(2) 1035-48

[17] Fan C and Zhou X 2010 Influence of operating conditions on the surface morphology of microcapsules prepared by in situ polymerization *Colloids and Surfaces A: Physicochemical and Engineering Aspects* **363**(1) 49-55

[18] Ma W, Zhang W, Zhao Y, Yu H 2016 Preparation and assessment of a self-healing material based on microcapsules filled with ethyl phenylacetate *J. Polymer Sci.* **133**(20)

[19] Pittala R K, Ben B S and Ben B A 2021 Self-healing performance assessment of epoxy resin and amine hardener encapsulated polymethyl methacrylate microcapsules reinforced epoxy composite *J. App. Polymer Sci.* **138**(23) 50550

[20] Khan N I, Haldera S, Goyat M S Effect of emulsifier on the properties of capsules for fabricating healing enabled next generation adhesive joints *XXX International Conference on Surface Modification Technologies (SMT30)* 29th June – 1st July 2016 Milan Italy

[21] Sun D, Chong Y B, Chen K and Yang J 2018 Chemically and thermally stable isocyanate microcapsules having good self-healing and self-lubricating performances *Chemical Engineering Journal* **346** 289-97

[22] Frihart C, Natesh A and Nagorny U 2001 Less odor and skinning with stabilized mercaptans for curing epoxies *Adhesives & Sealants Industry* **8**(1) 26-9

AUTHOR INDEX